Leading Science and Technology-Based Organizations

Leading Science and Technology-Based Organizations

Mastering the Fundamentals of Personal, Managerial, and Executive Leadership

Anthony Graffeo, Ph.D.

CRC Press
Taylor & Francis Group
Boca Raton London New York

CRC Press is an imprint of the
Taylor & Francis Group, an **informa** business

CRC Press
Taylor & Francis Group
6000 Broken Sound Parkway NW, Suite 300
Boca Raton, FL 33487-2742

Printed on acid-free paper

International Standard Book Number-13: 978-1-138-31080-3 (Hardback)
International Standard Book Number-13: 978-0-8153-9370-2 (Paperback)

Library of Congress Cataloging-in-Publication Data

Names: Graffeo, Anthony, author.
Title: Leading science and technology-based organizations : mastering the
fundamentals of personal, managerial, and executive leadership / Anthony Graffeo.
Other titles: Leading science and technology-based organizations
Description: Boca Raton : Taylor & Francis, a CRC title, part of the Taylor &
Francis imprint, a member of the Taylor & Francis Group, the academic
division of T&F Informa, plc, 2019. | Includes bibliographical references.
Identifiers: LCCN 2018013988| ISBN 9781138310803 (hardback : acid-free paper)
| ISBN 9780815393702 (paperback : acid-free paper) | ISBN 9781351188074 (e-book)
Subjects: LCSH: Science and technology intermediary
organizations—Management. | Executive coaching. | Employees—Coaching of. | Leadership.
Classification: LCC T175.5 .G73 2018 | DDC 658.4/092—dc23
LC record available at https://lccn.loc.gov/2018013988

Visit the Taylor & Francis Web site at
http://www.taylorandfrancis.com

and the CRC Press Web site at
http://www.crcpress.com

CONTENTS

Preface   xi
Author   xiii
Acknowledgments   xv
Introduction   xvii

1   The Performance Trilogy®: The Fundamental Processes of High Performance   1

Strategy—The Blueprint of Performance   3
Execution—Turning Intentions into Actions   4
Leadership—The Quarterback of Performance   5
Chapter Summary   6
    Strategy   7
    Execution   7
    Leadership   7
References   7

2   A Leadership Framework: Leading the Performance Trilogy®   9

Leading—Inspiring Faith in the Strategy   11
Is the Destination Desirable?   12
Is the Destination Achievable?   12
Is the Destination Beneficial?   13
Managing—Building Confidence in the Execution   14
Coaching—Gaining Trust through Development   16
Chapter Summary   18
References   19

PART I   PERSONAL LEADERSHIP: PERSONAL MASTERY THROUGH LIFELONG LEARNING AND SELF-DISCOVERY

3   Are You Ready to Lead?: Leadership Is Personal   23

Estimating the Magnitude of the Challenge   25
    Calculate the Degree of Difficulty of the Assignment   25

Estimate Your Level of Influence 26
Inventory Your Assigned Resources 27
What Are the Key Attributes You Bring to the Leadership Challenge? 27
Examine Your Motives 28
Discover Your True Talents 28
The Importance of Self-Awareness 29
What Is Self-Awareness? 30
Why Is Self-Awareness So Important? 31
Increasing Your Self-Awareness through Exploration and
Lifelong Learning 31
Chapter Summary 33
References 33

4   Leadership Attributes: Nine Essential Attributes That Ensure Success   35

Critical Leadership Attributes 35
Imagination—The Ability to Look at What Everyone Else Looks at, and
to See What No One Else Sees 36
Courage—The Ability to Act Despite Considerable Risks 37
Persuasiveness—The Ability to Influence 38
Critical Management Attributes 40
Conscientiousness—The Motivation and Discipline to Be Thorough and
Dependable 40
Productivity—The Ability to Manage One's Energy and Focus 41
Discernment—The Ability to Consistently Make Good Decisions
Based on Cognitive and Intuitive Judgment 43
Critical Coaching Attributes 44
Integrity—Behavior That Is Genuine in Words and Actions 45
Empathy—The Willingness to Walk a Mile in Another's Shoes 46
Teaching Skills—The Ability to Foster Self-Learning in Others 47
Chapter Summary 48
Leadership Attributes 48
Management Attributes 48
Coaching Attributes 48
References 49

PART II   MANAGERIAL LEADERSHIP: LEADING TEAMS

5   Transitioning from Me to We   53

I Am Excited about the Opportunity of Becoming a Manager 53
I Am Worried about the Challenge of Becoming a Manager 54
Managerial Roles, Responsibilities, and Expectations 55
Transitioning to a Management Role Will Be Both Exciting and Challenging 55
Developing a Managerial Identity and Style 57
Do Your Homework 60

Chapter Summary 61
References 62

6   **The Art of Supervision**   **63**

Take Control of Your Agenda 63
Get Organized 64
     Important/Urgent Activities 65
     Important/Nonurgent Activities 65
     Nonimportant/Urgent Activities 66
     Nonimportant/Nonurgent Activities 66
Focus Most of Your Attention on Your Staff 68
     Surround Yourself with Talented and Motivated Staff Aligned with
     Your Vision and Values 68
     Set High Performance Standards Starting with Yourself 70
     Critically, Honestly, and Frequently Evaluate Performance 72
     Coach up Your High Performers 73
     Marginalize Your Low Performers 75
Chapter Summary 76
References 77

7   **Project Leadership**   **79**

The Transition to Project Leadership 79
The Importance of Project Leadership 80
Technology-Based Business 82
     Business Development in a Technology-Based Business 82
     Staff Development in a Technology-Based Business 83
     Business Management in a Technology-Based Business 84
     The Customer Service Model 84
     What Is a Traditional Product-Driven Company? 86
     What Is a Market-Driven Company? 86
     It's Really a Continuum 87
     Most Companies Think They Are Market-Driven 87
     What Is the Real Difference between Being Product-Driven and
     Market-Driven? 87
     Technology Companies Are Moving toward Being Market-Driven 88
         Relationship Management 88
         Service Delivery 89
         Technology Development 89
         Implementing the Client-Service Model 89
         The Role and Responsibility of the Project Leader in Executing
         the Customer Service Model 90
         Project Leader's Role in Strategy 92
         Project Leader's Role in Execution 92
         Project Leader's Role in Development 94

Where the Project Leader Fits In   97
Chapter Summary   97
References   98

PART III   EXECUTIVE LEADERSHIP: LEADING ORGANIZATIONS

8   **Leading S&T-Based Organizations: Seven Critical Business Processes**   **101**

Strategic Planning   103
R&D Management   104
Financial Management   105
Business Development   106
Project Management   107
Product Development   108
Staff Development and Renewal   110
   Succession Planning   110
   Hiring and Promoting   111
   Coaching for Development and Performance   112
Chapter Summary   112
References   113

9   **Leading the Strategy: Setting the Direction and Developing the Road Map**   **115**

Strategic Planning Process   115
STEP 1. Self-Assessment—A Thorough Understanding of the Organization's
Strengths, Weaknesses, and Assets/Resources Available to It   118
   Product/Service   119
   Organizational Processes   119
   People   120
STEP 2. Client Feedback—Intelligence from Existing Clients on Their
Current Problems and Future Opportunities and the Value You Bring to
Their Organization   121
STEP 3. Competitor Analysis—The Strengths, Weaknesses, and Strategy of
Competitors' Products and Services Fighting for the Same Market Space   122
   Competitor's Product/Service   123
   Competitor's Organizational Processes   123
   Competitor's People   124
STEP 4. Technology Trends—An Awareness of Competing and Emerging
Technologies That Could Disrupt the Strategy   124
STEP 5. Market Analysis—A Thorough Knowledge of Market and
Industry Trends   126
STEP 6. Stakeholder Requirements—A Clear Understanding of Additional
Stakeholder Requirements Which Could Include Regulatory, Political,
and Social Factors That Could Positively or Negatively Affect the
Organization's License to Operate and Its Brand Image   129
STEP 7. Strategy Synthesis—Collecting and Synthesizing Disparate Data
and Information into Actionable Intelligence on Which to Make Decisions   131

Mission, Vision, and Values   132
Developing a Strategic Agenda   132
Market Strategy   134
Product Strategy   137
Resource Strategy   140
    A Staffing Strategy That Ensures Sufficient Depth of Talented and
    Motivated Staff   140
    A Financial Strategy That Maximizes the Use of Available Funds 141
    A Strategy to Acquire Expensive Facilities and Equipment 142
Regulatory Strategy   143
    Make the Regulatory Agency Your Partner, Not the Enemy 144
    A Regulatory Strategy Must Be Supported by Comprehensive and
    Unbiased Scientific Data Where the Potential Downside Risks Are
    Not Minimized   144
    Develop a Transparent Communications Strategy with the Public 145
Stakeholder Strategy   146
    Client Strategy   146
    Supplier Strategy   147
    Community Strategy   147
Chapter Summary   148
References   150

10

Managing the Execution: Translating Your Strategic Agenda into Actionable Objectives and Managing to Achieve Those Objectives

      151

Ensuring That the Strategy Is Rigorously Translated into Performance
Objectives That Are Actively Managed throughout the Organization   154
Developing Annual IOs That Ensure Meeting Your Strategic Thrusts   155
Establishing Key Performance Indicators and Milestones to Measure
whether the IOs Are Being Met   157
Cascading and Aligning the IOs to Every Level of the Organization   159
Conducting Monthly and Quarterly Review Meetings at Each Management
Level to Report on Progress versus Plan   160
  Strategic Performance Management   162
  Strategic Review Meetings   162
    Rationale   162
    Schedule   164
    Attendees   164
    Purpose   164
    Preparation and Inputs   164
    Agenda   164
    Expected Outcomes   165
    Follow-Up   165
    Best Practices   165
    Meeting Tools   166
  Operational Performance Management   169
  Operational Review Meetings   169
    Rationale   169

Schedule 170
Attendees 170
Purpose 170
Preparation and Inputs 170
Agenda 170
Expected Outcomes 171
Follow-Up 171
Best Practices 171
Meeting Tools 172
Communicating Any Progress Shortfalls and Actions to Be Taken to
Correct Them 174
Chapter Summary 175
Pitfalls 176
References 176

11 Coaching the Development: Coaching Is the Missing Ingredient in High
Performance 177
Staff Development and Renewal 179
Coaching for Development 180
Coaching for Performance 182
Balancing Organizational and Staff Needs—The True Meaning
of Alignment 184
Chapter Summary 185
References 185

12 Putting It All Together: Leading Is a Team Sport 187
Executing the Performance Trilogy® 187
Balancing Performance and Development 190
It's All about Talent Management 191
References 192

Index 193

PREFACE

As part of my responsibility as a vice president and managing director of two science and technology (S&T) organizations, I developed and taught several leadership courses that introduced scientists and engineers to the business of science. The critical role of S&T leaders is to create business and societal value from science. I felt it was important to augment their considerable technical skills by teaching them how to sell their ideas, manage their technical projects, and lead technical teams. Many have told me that they wished they had received this information and advice earlier in their careers. In my consulting practice, I have had the opportunity to introduce these leadership and management principles into several public and private sector S&T organizations with gratifying success. This book is a distillation of the key fundamentals of S&T leadership that I have learned as a student of the management literature and by trial and error in my 35 years of managing S&T organizations.

Why did I write this book? I have come to realize that there are three important phases in one's life. The learning phase is where society (parents, teachers, etc.) invests in you. The performance phase is where you return that investment and produce for society (marriage, children, and productive work). And finally, the teaching phase is where you pass on life's lessons to the next generation. This book represents my small contribution to this third phase in the hope that my lessons learned can shorten the leadership development process for those of you who practice the advice given.

While the literature on business leadership and management is extensive, including many outstanding contributions by CEOs, academics, and consultants, there is very little written on S&T leadership. In this book, written for S&T professionals and managers, I have attempted to translate fundamental leadership principles into language and examples that S&T professionals can better understand and appreciate.

I have tried to distill the best leadership and management principles and practices into a simple framework and language that can be easily understood and practiced by S&T professionals at all levels. While based on sound leadership theory, it is not an academic work, but a practical guide with anecdotes from real-life S&T situations. I have named this framework the Performance Trilogy®.

I developed a passion for leadership and management development midcareer from my mentor Dr. Bill Hitt, former Director of Management Development at Battelle, whom this book is dedicated. It is my hope that reading this book will stimulate younger scientists and engineers to further explore and consider taking on leadership assignments and even a profession in S&T management. Also, for current S&T managers to rededicate their careers to creating value from science and communicating the important contributions that science is making to economic development and quality of life that we all currently enjoy.

We enjoy the fruits of S&T every day of our lives. However, there is a major movement afoot to view S&T in a bad light. This can hamper future S&T discoveries and applications for the greater good of society. We need more S&T leaders who can not only create and manage science but can also communicate the advantages and explain the risks of scientific discoveries in a rational way to the general public.

As Bronowski so elegantly described in his book, *Science and Human Values*, "The world today is made, it is powered by science; and for any man to abdicate an interest in science is to walk with open eyes towards slavery."

For further information concerning the concepts presented in this book and training opportunities, please visit my website at www.graffeoandassociates.com or feel free to email me at Tony.graffeo@graffeoandassociates.com

☐ Reference

1. Matheson, David, Matheson, Jim, *The Smart Organization, Creating Value through Strategic R&D*, Harvard Business School Press, Boston, MA, 1998.

AUTHOR

Dr. Graffeo's breadth and depth of experience in managing S&T over a 40-year career makes him uniquely qualified to teach S&T leadership. He has taught, consulted with, and coached more than 1,000 scientists and engineers from three continents on managing S&T and currently serves as a coach for research and development (R&D) executives. He was appointed as a professor at Northeastern University in 2017 and is teaching graduate courses in Biotechnology Entrepreneurship and Leadership.

He started his professional journey as a cooperative educational student at Northeastern University, serving as a laboratory bench scientist for several Boston technology companies. He went on to receive a Ph.D. in bioanalytical chemistry from Northeastern and was hired by Battelle Memorial Institute to build a laboratory capability in high-performance liquid chromatography.

After serving as a bench scientist for 2 years, he was promoted as a technical group leader and associate manager: the youngest in the history of the Institute. For the next 10 years, he took on successively more responsibility as chemistry section manager and department manager, and laboratory director of Battelle's Marine Science laboratory. He gained a reputation as a "turnaround specialist" with a successful track record of fixing troubled S&T programs and growing them into recognized centers of excellence. At Battelle, he gained valuable experience in the business of contract research, selling and managing research to the U.S. government.

Arthur D. Little Inc. (ADL) then recruited him to head up two of their business segments as vice president and managing director of Life Sciences and Environmental S&T. At ADL, he managed a technical and consulting staff of 150 scientists and engineers. In the Life Sciences business segment, he had profit and loss responsibility for toxicology, pharmacology, metabolism and pharmacokinetics, product formulation, and analytical chemistry. In the Environmental S&T business segment, he was responsible for environmental technology, pollution prevention, environmental monitoring and assessment, bioremediation, oil spill prevention and response, and environmental forensics. At ADL, he gained valuable experience in consulting with major industry segments: pharmaceutical, biotechnology, and oil and gas industry.

He then spun out the life sciences business from ADL with the help of an angel investor and formed Biodevelopment Laboratories to serve the emerging biotech industry. Within 3 years, Biodevelopment Laboratories had built a thriving business with the pharmaceutical and biotech industry, resulting in its acquisition by Genzyme for a 300%

return on investment. In spinning out Biodevelopment Laboratories, he gained valuable entrepreneurial experience in creating and financing an S&T business.

He then returned to Battelle to head up its international operations in Latin America and Europe. He established Battelle's Mexico operations and had corporate responsibility for Battelle's Environmental and Agricultural Product Registration laboratory operations in Geneva, Switzerland and in Onger, UK. He expanded the UK operations with the acquisition of an agrochemical synthesis laboratory. He had corporate marketing responsibility for the Middle East, Europe, and Latin America. In his second stay at Battelle, he gained valuable experience in the cultural and regulatory nuances of international business and the acquisition process.

Upon his retirement from Battelle in 2007, he was asked to write a report on the findings of a Kuwaiti Blue-Ribbon Panel commissioned by the Amir to develop a Science, Technology, and Innovation agenda for the country. This led to helping the Kuwait Institute for Scientific Research and Kuwait Foundation for the Advancement of Science to implement the Blue-Ribbon Panel's recommendations and develop a transformational science, technology, and innovation strategy for the country.

Dr. Graffeo is currently a founder and president of Graffeo & Associates, an organization focused on helping S&T organizations create winning strategies, improve their performance, and develop their management leadership. His accomplishments include

- As a project and program manager, he led several multimillion-dollar government programs with the US Department of Defense, Environmental Protection Agency, and Health and Human Services in energy, environment, and biopharma.

- As a business unit manager, he turned around several troubled S&T organizations in both United States and Europe and grew them into profitable, recognized centers of excellence.

- As a global business manager, he developed business strategies, conducted international technology assessments, acquired and integrated companies, and developed several strategic client relationships.

- As an entrepreneur, he spun out a contract research organization from ADL, managed its growth, and sold it for three times its acquisition price in 3 years.

As a teacher and coach, he developed a unique ability to translate complex business and management concepts into a language more easily understood by scientists and engineers. He has developed a comprehensive set of leadership-training programs for S&T managers to help them critically evaluate their business processes, the value of their technology portfolios and the quality and depth of their leadership.

ACKNOWLEDGMENTS

I would like to acknowledge the following important people in my life without which I could not have written this book. To my mentor the late Bill Hitt, Leadership Development Manager at Battelle, who inspired me to take up a career in management and guided me for many years. My current mission is to follow in his footsteps. To two of my most influential managers, Richard Nathan and Gabe Kovacs, who allowed me the freedom to experiment and flourish while gently coaching me without me even knowing it. A lot of the wisdom in this book I learned from them. To two of my associates, Jon Olson and Richard Chidester, with whom I worked closely for the past 10 years, and who have helped me refine my thinking about leadership training and development. To my son Michael, who has continuously encouraged me to pursue my passion for teaching and I am proud to say is following in his dad's footsteps but with bigger shoes. And last but not least, my wonderful wife Lin who has supported me throughout this writing process and has spent too many days alone while I was writing. She is the main reason why I wake up every morning a very happy man.

INTRODUCTION

Although there have been thousands of books written on organizational management and leadership, very few have specifically addressed the unique challenge of leading an S&T organization. The few books written on the subject have focused exclusively on strategy or policy [1,2] or the interpersonal relationships of scientists [3].

Leading Science and Technology-Based Organizations will be the first book written by a practicing S&T leader that presents a practical framework for leading, managing, and coaching throughout one's career. Best practices for each of the leadership fundamentals and organizational processes will be highlighted and lessons learned will be discussed based on more than 40 years of experience in managing global S&T organizations and teaching S&T leadership to more than 1,000 scientists and engineers in the United States, Latin America, Europe, and the Middle East.

The challenges and insights that I have learned over the years have come directly from trying to lead and manage scientists and engineers, sometimes successfully, sometimes not. This experience has led me to believe that if you can lead scientists and engineers, you can lead anybody. I mean this as a compliment as technically trained staff cannot be fooled by superficial or authoritarian managers. They are highly intelligent and analytical and can quickly point out the flaws in a manager's approach or intentions. They are mostly devoid of business knowledge and oftentimes are the chief cynics of management initiatives. By the nature of their training, they resist teamwork and prefer total control over their work assignments without any interference from management.

Solving these challenges has helped me become a better leader and manager to the point now where I believe that the principles taught in the book are extremely robust and can be used to manage any team or organization.

The approach in the book is based on simple, practical advice on management fundamentals. I have interacted with many S&T leaders who have attended several management courses and yet still underperform. In coaching some of these leaders, I have come to the conclusion that they have heard and understood that the principles of good management are oftentimes from reputable international sources. What they are missing is the "how to," unless they have been fortunate to have worked under a good leader who has coached them to make sure that they know how to do their job.

After an introduction, this book is organized into three parts paralleling the leadership career path of S&T professionals. In the Introduction, the Performance Trilogy is introduced (Chapter 1) and a leadership framework is described (Chapter 2) that underpins the remainder of the book.

In Part 1, the focus is on personal leadership. The hardest person you will ever have to lead is yourself. The lessons learned through increased self-awareness and self-development form the foundation of your leadership of others. Great leaders understand the importance of leading by example and pursuing personal mastery through lifelong learning.

In Part 2, personal leadership skills are leveraged, and impact is increased through managerial leadership. One of the most difficult career transitions that scientists and engineers have to make is going from an individual performer to a manager; i.e., a mindset change from "me" to "we." Throughout middle management, your leadership skills will be challenged, and your success will depend on the degree of maturity of your personal leadership skills learned in Part 1.

In Part 3, an additional career transition takes place from managerial leadership to executive leadership. Executive leadership requires the understanding and oversight of the critical business processes of the organization and the reliance on the functional expertise of others in making decisions. Your executive leadership success will depend on identifying and developing the leaders of these business processes.

☐ **References**

1. Matheson, David, Matheson, Jim, *The Smart Organization, Creating Value through Strategic R&D*, Harvard Business School Press, Boston, MA, 1998.
2. Millar, William L., Morris, Langdon, *Fourth Generation R&D*, John Wiley, 1999.
3. Cohen, Carl M., Cohen, Suzanne L., *Lab Dynamics*, Cold Spring Harbor Laboratory Press, Cold Spring Harbor, NY, 2012.

The Performance Trilogy®
The Fundamental Processes of High Performance

The degree of complexity in today's science and technology (S&T) organizations has reached the point that even experienced technical managers are having a hard time focusing on the critical activities that lead to high performance and successful results. Leadership is one of the most studied subjects in the world and the least understood one [1]. The business literature today provides an overwhelming array of leadership and management articles and books focused on achieving high performance. This literature is not as helpful as it could be. On the one hand, they are written for fellow experts and introduce complex and oftentimes confusing theories and language to describe management processes. On the other hand, there are many articles and books written for novices, describing quick management fixes and magic bullets to solve all their problems. This book is not about new theories or quick fixes but presents a point of view that there are three fundamental processes that must be mastered to attain outstanding and sustainable personal, management, and executive performance. I have labeled these three processes as the Performance Trilogy®.

Earlier in my career, I received some sage advice from one of my mentors that had a dramatic effect on my performance as a manager and leader. He said to me "whenever I reviewed a failed initiative in our company, it invariably was due to one of three factors: a faulty strategy, poor execution, or the wrong leadership." Since then, I focused my attention on these three fundamental processes of strategy, execution, and leadership whenever I took on new assignments. Throughout my career, it helped me turn around four different operating divisions on three continents that had lost their technological edge and were losing money. In all cases, my diagnosis identified a problem with one or more of these three fundamental processes: strategy, execution, and leadership. Much later in my career, I managed to read *Execution—The Discipline of Getting Things Done* by Larry Bossidy and Rahm Charan [2], which validated almost everything I was practicing as a manager and a leader. It was also good to know that the fundamental principles I learned and practiced at work were being taught at General Electric's Leadership Development Center in Crotonville, ranked Number 1 in the world in Leadership Development [3].

In one turnaround assignment, I discovered that the strategy was clearly faulty. Markets had shifted dramatically, and the organization was wedded to the past, missing new opportunities in an emerging market. In two other management assignments, the loss of several clients and the resulting financial losses were incorrectly attributed to a high-cost structure, when in reality the project execution was not meeting client expectations. In another assignment, the key leader had left the organization, and there was no leadership backup or depth to continue a successful operation.

1.1

This book is not about new theories or quick fixes, but presents a point of view that there are leadership fundamentals that must be mastered to truly attain sustainable high performance: I have labeled these as the Performance Trilogy

In my first turnaround attempt, it took about 3 years to return the organization to profitability and develop a stable client base. The first year was spent diagnosing the problems and evaluating the existing leadership team. Armed with this information, the strategy was revised, execution more closely managed, and the right leadership was put in place in year two. We saw dramatically improved results in year three. While successful, there were many lessons to be learned. I found that the more opportunities I had to practice the Performance Trilogy fundamentals, the better and faster I got over time (Figure 1.1).

From my perspective, there are only three fundamental processes that leaders have to get right to ensure high performance:

1. Developing a winning strategy and getting buy-in from those who need to implement it

2. Ensuring that the strategy is rigorously executed by translating key strategic thrusts into performance objectives that are actively managed throughout the organization

3. Selecting and developing talented and motivated leaders to execute the strategy.

FIGURE 1.1 The Performance Trilogy.

Most business failures can be attributed to these three factors: a faulty strategy; poor execution; or the wrong person leading its implementation. By focusing on these three fundamentals, all the other management activities can be viewed in context and better prioritized.

In principle, this all sounds pretty simple, right? If it is so simple, why is it that it is so rarely practiced? Well, because it's easier said than done. There is a myriad of everyday distractions that constantly interfere with you and your team's ability to keep your eye on the target. There is also a great deal of self-deception that rationalizes behavior that is counterproductive to high performance. By diligently practicing the fundamentals of the Performance Trilogy, you can avoid common activity traps and faulty reasoning and focus your time on the critical success factors that ensure high performance. It's time to stop looking for quick fixes: get real and start practicing the fundamentals.

☐ Strategy—The Blueprint of Performance

Strategy is truly the blueprint of performance, and I have found that most strategic plans are not compelling enough. When leading an S&T organization, research and development (R&D) decisions have tremendous leverage as they sit at the beginning of the entire value chain. These decisions are particularly difficult because of the many uncertainties.

1.2

If you had unlimited time, money, and resources, strategy would become less important.

1. The time between the decision point and the point at which the cash register starts ringing is typically very long and filled with uncertainty.

2. The R&D process is inherently uncertain (without uncertainty, there would be no R&D). It is difficult to predict the success rate of most R&D projects.

3. The markets to be served are most uncertain at the time R&D projects are commissioned.

4. Successful R&D often takes a company into unfamiliar areas requiring partnerships, alliances, or acquisitions and new ways of doing business [4].

Most strategic plans suffer from ignorance, arrogance, or both. Oftentimes, plans are written with minimal information about the real needs and desires of clients and team members. Worse still, critical resources and team capabilities are often overestimated and the competition underestimated, leading to failed strategies right from the start. It is important to remember that an excellent strategy will always be beaten by an outstanding one. On the other hand, a good strategy will beat a mediocre one. The key is to develop a winning strategy, one that will beat your competition.

Strategic planning involves developing a vision (where you want to go), a road map (how you are going to get there), and a resource plan (identifying all of the resources you need at the right time and the right place). It takes into account all of the potential barriers and resource constraints you can think of that you might encounter and develop alternative scenarios that potentially can overcome those barriers. You can obtain generic strategic plans from any good management consulting firm, but the content that goes into your plan must come from those knowledgeable about your capabilities, clients, and competitors.

The operative definition of strategy is the allocation of scarce resources. The reason why strategy is the first fundamental process and is so critical to performance is that it narrows down the myriad number of potential activities to a manageable few. You and your team need to focus only on those activities that will lead to achieving your vision. If you had unlimited time, money, and resources, a strategy would be less important for you as it could cover many options and perform multiple essential and nonessential activities until you reached your destination or goal. Since resources in the real world are finite, they must be carefully focused on essential activities that produce results. The larger the team gets and the more layers of the organization, the more difficult it becomes to focus everyone's attention. A compelling strategy provides clarity and focus.

1.3

Nothing feels better than to lead people to excellence, fulfillment, and collective achievement.

☐ Execution—Turning Intentions into Actions

Once a compelling strategy is developed, the second process of the Performance Trilogy involves execution, i.e., translating your strategy into actionable objectives and managing to achieve those objectives. This is easier said than done. The creative work of strategy development suddenly turns into hard work. In my experience, I have rarely seen this translation done very well for two reasons. First, individual team members have personal aspirations and agendas that do not align 100% with the proposed organizational strategy and tend to go off on tangents. Second, senior leaders incorrectly assume that this translation and management deals with operations and not strategy and delegate it to their subordinates. In both cases, manager's time, the scarcest resource in most organizations, is not being fully used to advance those critical objectives that would ensure success.

Both the development of the strategy and translating it into an execution plan with actionable objectives must be done jointly between you and your team members. As discussed in the next chapter on leading the Performance Trilogy, obtaining not only buy-in but also enthusiasm from team members is critical to success. Objectives should simultaneously support the strategy as well as the personal goals of each individual. This is a good definition of "alignment."

Each of the team member's objectives must be directly linked to the advancement of the strategy such that the output of each objective results in a desired strategic outcome. This requires practice and experience to do well. The minimum requirement should be at least a 70% alignment (i.e. 70% of the objective's output contributes to the strategic outcome). Any percentage lower than that would require a serious reexamination of that team member's suitability for and/or willingness to take on the assignment.

Unfortunately, too many managers view setting performance objectives as a necessary evil and an undesirable chore. After reading this book, I hope that it will increase the awareness of the importance of performance planning and review as a fundamental element in the execution process. In Chapter 7, I have developed a codified active management process to support your subordinates' development while ensuring organizational performance. By focusing both on the development planning and review process (DPR) as well as the performance planning and review process (PPR), I believe that this novel approach will finally make the review process not only more effective but also pleasurable! It will require a change in your mindset and some success, but I assure you that nothing feels better than to lead people to excellence, fulfillment, and collective achievement [5]. Most of us are aware of this after one of our sports teams wins a championship.

> **1.4**
>
> There is a common myth that is widely accepted as fact that a senior manager's job is to focus on strategy and not get his or her "hands dirty" with operational issues. Nothing could be further from the truth.

Once the strategy is developed and an execution plan put in place with actionable objectives, the senior leader needs to take off his leadership hat and put on his management hat and ensure that team members stay on track. There is a common myth that is widely accepted as fact that a senior manager's job is to focus on strategy and not get his or her "hands dirty" with operational issues. Nothing could be further from the truth.

Many a good strategy has failed from poor execution due to senior managers (I call them macromanagers) who "abandon" their direct reports once performance objectives are produced. It is the responsibility of the senior manager and architect of the strategy to ensure not only that the strategy gets translated into actionable objectives but also that the performance is managed. This is accomplished not by macromanaging or micromanaging but by active management and regular governance of the progress of objectives. This involves supporting each team member by removing organizational barriers, identifying performance shortfalls, determining lessons learned, and correcting deficiencies. In the upcoming section on leadership, the execution process will be discussed in more detail.

☐ Leadership—The Quarterback of Performance

Developing a successful strategy and executing it flawlessly is totally dependent on the selection and development of the team leader and team members chosen to perform the work. Using a football analogy, they represent the first string led by the quarterback. High performance is directly related to their talent level and motivation. The two biggest mistakes a manager can make is selecting the wrong leader for the job and tolerating underperformance.

In selecting team members to help develop and execute strategy, the main focus should be on talent, motivation, and experience in that order. Too many managers weigh experience too heavily at their own peril. In my experience, talent is by far the most important criteria for selection, and it should closely match the job requirements. Too often, even experienced staff are placed in assignments that don't match their unique talents, resulting in underperformance and decreased motivation.

Second in importance is motivation and enthusiasm for the job. Enthusiasm can overcome many deficiencies in experience as well as barriers to performance that exist in most organizations. This is particularly important when the team members are close to 100%, aligned with the organizational strategy and execution plan.

Finally, experience does count, provided it is based on results and not activities or time on the job. For example, you have to determine whether the candidate has 10 years' experience or one year's experience ten times. The best area to focus on is the candidate's track record of performance at achieving results that were asked of him.

Once the right candidate is selected as a member of your team and has bought into the strategy and is aligned with the execution plan, your role becomes much easier. Your main focus now becomes a continuous development of that team member to achieve his career goals, since they are essentially aligned with the organization's goals.

We all know that the rate of technological progress is close to exponential, and as such, it is very easy to fall behind without continuous development. For example, if you had to conduct a new assignment with a team member that was brought to the present time from the 1990s, he or she would have a great deal of difficulty succeeding. Technological change is such that knowledge worker productivity has increased by an order of magnitude over the past 15 years. The speed at which we are able to collect, process, file, and retrieve data today would astonish workers from the 1990s.

1.5

High performance depends on talented and motivated staff, and a well-designed and executed development program increases both.

By development, I mean the continuous improvement of a person's awareness, skills, and attitudes that will ultimately contribute to higher performance. The leadership development process is the most talked about and least well practiced. Too often, development plans are an afterthought in the performance planning process, not taken seriously enough, and rarely aligned with actionable objectives and expected future performance. Too many managers view development as a cost rather than an investment, despite the proven fact that it has a very high return on investment. High performance depends on skilled and motivated staff, and a well-designed and executed development program increases both. By adopting the strategy in this book using the DPR and PPR processes in that order, you will be light years ahead of most organizations in ensuring high performance.

I was impressed with a story that was passed on to me by one of my associates of a discussion between the CEO of a company and the head of human resources (HR). The HR manager was questioning the amount of budget allocated for training. He asked, "what if we invest all this money and the staff decide to leave"? The CEO's response was "what if we don't invest in their training and they decide to stay"?

The Performance Trilogy is about fundamentals: the development and execution of a winning strategy by a talented, aligned, and motivated team. Clearly, this doesn't happen without extraordinary leadership, which is the topic of the next chapter.

☐ Chapter Summary

The fundamentals of the Performance Trilogy:

1. Developing a winning strategy and getting buy-in from those who need to implement it

2. Ensuring that the strategy is rigorously translated into performance objectives that are actively managed throughout the organization

3. Selecting and developing talented and motivated leaders to execute the strategy.

Strategy

Developing a vision (where you want to go), a road map (how you are going to get there), and a resource plan (identifying all of the resources you need at the right time and the right place).

The objective is not to develop an excellent strategy but a winning one.

Execution

Each of the team member's objectives must be directly linked to the advancement of the strategy, such that the output of each objective results in a desired strategic outcome.

Effective execution is accomplished by active management: focusing both on the DPR as well as the PPR.

Leadership

In selecting team members to help develop and execute strategy, the main focus should be on talent, motivation, and experience in that order.

Determine whether the chosen candidate for your team has ten years' experience or one year's experience ten times.

High performance depends on skilled and motivated staff, and a well-designed and executed development program increases both.

☐ References

1. Adair, John, *The Inspirational Leader*, Kogan Page, London, 2003.
2. Bossidy, Larry, Charan, Ram, *Execution- The Discipline of Getting Things Done*, Crown Business, New York, 2002.
3. The Aon Hewitt Top Companies for Leaders list.
4. Matheson, David, Matheson, Jim, *The Smart Organization, Creating Value through Strategic R&D*, Harvard Business School Press, Boston, MA, 1998.
5. McKee, Annie, Boyatzis, Richard, Johnston, Frances, *Becoming a Resonant Leader*, Harvard Business Press, Boston, MA, 2008.

CHAPTER

A Leadership Framework
Leading the Performance Trilogy®

In the previous chapter on the Performance Trilogy®, the path to high performance was to focus on three fundamental processes: strategy, execution, and leadership. Understanding each of the processes and utilizing certain process tools, however, is necessary but not sufficient for success. Specific leadership roles, responsibilities, and actions are required during each of these three processes to inspire team members and keep them committed. If team members are disengaged, you need to look in the mirror and take ownership of the problem.

In this chapter, a leadership framework is introduced focused on the specific roles needed and actions you must take as a leader to generate emotions that can drive team members to want to follow you. As a leader you must inspire faith in your team in your strategy, build confidence in the execution of the strategy, and engender trust among team members in your motives. One of the biggest hurdles that scientists and engineers must overcome is their lack of awareness of the importance of these emotions in building their teams and managing performance. In the following pages, I hope to present an approach to leading the Performance Trilogy to stimulate these emotions based on logic backed up by data.

2.1

Leadership is the art and science of creating the future by inspiring others to follow you to your desired destination, building their confidence in the path taken, and gaining their trust that you have their interests at heart. The essence of leadership is the ability to transform ideas into significant results.

Leadership is a concept that everyone knows when he or she sees it, but that few really understand the complexity or "formula" for leadership. If you ask ten people to define leadership, you will get ten different answers. If you search the word leadership in Google, you will get over 100 million hits. Such a plethora of perspectives gives the impression that the concept of leadership is well defined and understood. However, whenever I have asked participants in my leadership and management courses about their understanding of the term leadership, I have gotten vague answers that vary widely based on their own specific experiences. My sample size is over a thousand senior scientists, engineers, and technical managers taken over a period of 25 years across multiple cultures in the United States, Latin America, Europe, and the Middle East.

I am confident that the generally accepted views on leadership have been distorted by myths that have been built up over the years. Thousands of books have been written about great political, religious, military, and business leaders that give the impression that leaders are born not made and that leadership is reserved for the very few best, brightest, and charismatic. Many management and leadership models have been proposed, but few describe what leaders actually do and how they make decisions. Popular books and articles are published continuously that promise quick fixes and magic bullets on management and leadership. Fortunately, there is an emerging point of view from academic experts that leadership is important at all levels of the organization (distributed leadership) and not just for managers.

Of the hundreds of books that I have read on management and leadership, I decided to list my top ten in this book for those of you who wanted to build on your personal experience as I did. The task proved impossible as there are so many must reads. With great difficulty, I was able to winnow it down to the top 20 [1–20], and at the end of the chapter, I forced myself to list the top ten that influenced me the most.

These prevailing myths along with the lack of good role models have prevented many science and technology (S&T) managers and senior technical staff without business degrees from thinking of themselves as team or organizational leaders. As a result, many with high potential have hesitated to take appropriate steps to develop their leadership skills and take on challenging management assignments. Unfortunately, many more have been unsuccessfully thrust into management leadership roles without the proper understanding, perspective, or training. In my experience, it is extremely rare to find S&T managers who consistently practice the leadership concepts presented in this book.

I believe that everyone is born with some leadership talent and can develop additional skills that will allow them to successfully execute the Performance Trilogy. Webster defines a leader as "one who guides." The management literature defines leaders as those people who achieve goals. Each of these definitions captures elements of leadership but provide no guidance on what to do in planning and executing a leadership initiative or how to elicit feedback on how well you are doing. I offer the following definition and resulting model based on over 35 years of personal successes and failures in leading and coaching scientists and engineers.

> Leadership is the art and science of creating the future by inspiring others to follow you to your desired destination, building their confidence in the path taken, and gaining their trust that you have their interests at heart. The essence of leadership is the ability to transform ideas into significant results.

While this definition seems straightforward and perhaps simplistic, as you shall see, even if you are skilled enough to create a compelling strategy, it takes a great deal of self-awareness, skill development, and hard work to successfully lead it through to completion. There are no magic bullets. The leading expert on leadership, John Kotter [20], has suggested that 70% of all corporate strategies fail to achieve their potential. I am confident that by adopting the leadership framework in this chapter, you will dramatically increase the odds of success.

Leading each of the three processes of the Performance Trilogy requires that you assume different leadership roles and responsibilities. During the strategy process, your role is to "lead from the front" to create and promote a compelling strategy that inspires faith among your team. This is the most traditional view of term "leadership." During the execution phase, your role is to "lead from the middle" to build confidence among your team with your ability to execute the strategy to produce results. This is the traditional view of "management." Throughout the leadership process, your role is to "lead from the rear" to gain your team's trust by empowering them, guiding them and helping them grow as leaders. This is the traditional view of "coaching." Ultimately, in my mind asking whether someone is a leader or a manager is the wrong question. Using the Performance Trilogy Leadership Framework, successful and sustainable leadership involves all three roles: leading, managing, and coaching for every initiative.

2.2

Leadership = leading strategy + managing execution + coaching development

☐ Leading—Inspiring Faith in the Strategy

As discussed in Chapter 1, the first process to be mastered in the Performance Trilogy is the development of a winning strategy. A winning strategy must be compelling, that is, well thought out, highly competitive, and well communicated. More than likely, you will be asking your team to make substantial changes in their workflow and current roles and responsibilities. It will be obvious that they will be required to take on considerable professional risk based on your word that the strategy will be successful. This will require considerable faith by your team that you can deliver on your promise. *By faith, I mean believing in something, which you cannot prove.* This represents a substantial hurdle when trying to lead a technical team of scientists and engineers, who value decisions based on data not speculation. It's no wonder that scientists are usually the biggest skeptics of new management initiatives. Therefore, it is especially important to provide a sound hypothesis for your vision and be willing to have it rigorously challenged if you expect your technical team to share your vision. Only when your team has faith in your strategy and a "shared vision" of the destination can you confidently proceed to the execution phase.

2.3

To inspire your team to leave their comfort zone and commit to the new strategy, you need to explain why achieving the vision will be important, meaningful, and worthwhile.

This is why most successful leaders involve their team members from the very beginning in the development of a strategy. When team members have a "seat at the table" and an opportunity to review the supporting data, challenge the assumptions, and have input, they are more likely to own and defend the strategy. Too often, potential leaders make the mistake of moving forward prematurely with skeptical team members and then wonder why they fail in executing the strategy, often blaming their team rather than themselves. When you and your team develop the strategy, you should ask yourself three critical questions and have solid answers before moving on to execution.

☐ Is the Destination Desirable?

It is important that the destination you choose (or vision) be highly desirable. To get inspired by any vision, be it a personal, team, or institutional one, the vision needs to provide meaning and purpose not only to you and your superiors but also to the team that needs to execute. So, it is not only important to describe "what" the ultimate destination is, but more importantly "why" it is important and urgent enough that it is worth pursuing. The lack of a sense of urgency and shared vision among team members are two of the major causes of failed strategies. To inspire your team to leave their comfort zone and commit to a new strategy, you need to explain why achieving the vision will be important, meaningful, and worthwhile.

Human nature tells us that there are basically two major reasons why people are willing to make substantial changes in life: the first is through inspiration and the second is through desperation. Examples abound of people who have been inspired to take up causes due to the death of a family member or social injustice. While the meaning and purpose of your business vision may not be as dramatic, the more thought that goes into its meaning and purpose, the more willing your team will be to embrace the changes that may be necessary to get there. I encourage you to view the TED Talk by Simon Sinek to see the power of "why" [21].

Alternatively, although not quite as impactful, you may need to point out what the consequences might be if you don't pursue the strategy. Examples of major changes caused by desperation include quitting smoking after a heart attack and using seat belts after seeing the consequences of a major accident. Once again, while the downside of not taking action on your initiative may not be quite as dramatic, the loss of independence, resources, pay raises, bonuses, or even job losses may influence your team to make the necessary changes to implement a new vision.

No one wants to spend their life working on meaningless chores. To generate enthusiasm for your vision, you must find meaning for everyone on the team. To illustrate, I am reminded of the story of a senior construction manager who decided to spend a day among his various crews to find out why some were more productive than others. He went to a bricklayer from the first crew, which was the least productive, and asked, "what is your job function?" The bricklayer said "my job is to place one brick on top of the wall and cement it in place. I continue to do that all day, with a morning and afternoon break, then I get to go home." When asked the same question from a bricklayer on the second crew that was twice as productive as the first, he got the following answer. "Our job as a team is to build this brick wall as fast as we can within quality specifications to beat the other teams. If we do, at the end of the week we get a bonus." The third crew was the most productive, and when asked the same question responded, "we have the privilege of helping to build this magnificent cathedral." This simple example illustrates the importance of vision and explaining the "why" as well as the "what" in generating enthusiasm and productivity. Are you on a team that is putting one brick on top of another or are you building a cathedral?

☐ Is the Destination Achievable?

Once your team is convinced that the destination is highly desirable, you then need to convince them that the path chosen (strategy) has a reasonable chance of getting you there. Asking team members to accept and follow your strategy oftentimes involves considerable risks. To ask them to leave their comfort zones and take on such risks requires you to prove to them that your strategy is well thought out and you and the team have the talent, motivation, skills, and knowledge necessary to successfully execute the strategy. The strategy needs to be highly specific, including a road map and an

---

2.4

Oftentimes, bright and talented potential leaders make critical mistakes in developing strategy due to ignorance, arrogance, or both.

---

execution plan detailing each of the required steps. Potential barriers need to be identified and actions identified to overcome them.

Oftentimes, bright and talented potential leaders make critical mistakes in developing strategy due to ignorance, arrogance, or both. Ignorance is displayed most often by a lack of understanding of client or stakeholder needs and desires. Too often assumptions are made without obtaining the information directly from the clients or stakeholders. Also, competitor's strengths are often underestimated due to lack of information. Arrogance is often displayed in overestimating the strengths of the team as well as the resources that might be available within the organization. Even worse, valuable information is ignored because it doesn't support the strategy.

A common misconception is that you must show absolute confidence in your strategy in order to convince the skeptics. This doesn't work with highly intelligent technical staff as you cannot hide the potential flaws of your strategy through rhetoric. It is important to be willing to have your strategy challenged by your team members. The more questions that you can answer to their satisfaction, the more confidence they will develop in you. Questions that you cannot answer will point out potential weaknesses to the strategy that need to be shored up through more information or backup plans. Rigorous examination by your team as well as trusted advisors will strengthen the strategy, gain the confidence of team members, and increase the probability of success.

How often have you seen such openness and transparency in your organization? As I have stated, knowing the fundamentals is not good enough. You must be willing to practice them even if it proves that you don't have all the answers.

☐ Is the Destination Beneficial?

If you have answered the first two questions well, the majority of your team members are beginning to overcome their skepticism and believe that the destination is indeed desirable (shared vision) and that the path chosen (strategy) has a reasonable chance of succeeding. Too many potential leaders assume that strategy development is completed at this point and move on to execution. They never get to the third question

---

2.5

Spending time with each of your key team members to determine what their aspirations are and how this assignment may fulfill them may be the single most important activity that you can conduct during the strategy phase.

---

that needs to be answered: how will the initiative, if successful, benefit each of the team members?

As the leader, you are acutely aware of how success will benefit you and the credit that will accrue to you as the creator and leader. But each of the team members have their own desires about what they wish to gain from providing their support. It always amazes me how often leaders just guess and misread their team's needs. It is always worth remembering that people do things for their reasons not yours.

Meeting corporate goals is not often on the top of the list for most scientists and engineers. Spending time with each of your key team members to determine what their aspirations are and how this assignment may fulfill them may be the single most important activity that you can conduct during the strategy phase. This is the only way to win over their hearts and minds. Explore both extrinsic and intrinsic rewards that your team members are expecting and make realistic commitments to them based on the performance you expect. Also discuss potential risks that they may be taking and ways to mitigate them. If your goals and that of a team members' are not aligned, it is time to discuss putting him on a different team. This will benefit both of you.

Only when you are confident that you have successfully answered all three questions in the minds of each of your key team members should you proceed to the execution phase. For organizational initiatives, you and your team will need to develop a more formal communication plan where these three questions are openly discussed among the wider audience. This is easier said than done. Suffice it to say that, in my 40 years of organizational life, I've never once seen a strategy overcommunicated!

☐ Managing—Building Confidence in the Execution

Once you and your team are convinced that you have developed a compelling strategy and communicated it to the staff, you can turn your attention to the execution process. The first step is to translate the strategy into actionable objectives for you and your team that will lead to the desired strategic outcomes. For an organizational initia-

> **2.6**
> Management takes hard work, sweat, and practice. Plans are only good intentions unless they are immediately followed by hard work.

tive, the objectives need to be cascaded down through each management level of the organization. Each of the actionable objectives needs to harness and direct the available resources toward achieving the strategic outcomes. This is easier said than done. In my experience in auditing hundreds of performance plans in several organizations, I have rarely seen direct alignment between the actionable objectives of team members with strategic outcomes. In many cases, team member activities, however well-intentioned, do not lead to the intended results of advancing the strategy.

Unfortunately, too often strategies are not translated at all into actionable objectives, and strategic plans become no more than annual exercises that get left on the shelf. Peter Drucker [8], the leading expert in management consulting, points out that knowledge, wisdom, and expertise are useless without action. The most important part of this stage of execution is to develop confidence early with team members by developing actionable objectives with set milestones. Successes achieved early in the process can set the right tone for the entire strategy. Ideas are no more than unrealized dreams; strategies make ideas possible; developing and scheduling actionable objectives make ideas real.

The second step is to manage the execution process by following up with *regular* reviews of your team's actual performance. It is important to recognize that the actionable objectives are just good intentions on a piece of paper and not a commitment. Too many technical staff view the role of managers as just administrators and nothing could be further from the truth. While good administration is important, managing is an active verb, and performance management is a discipline that requires not only setting actionable objectives but also consistently following up to make sure that the objectives are being met.

There are two mistakes that I see all the time with the implementation of the discipline of performance management. The first one is to micromanage team members. The difference between active management and micromanagement is that an active manager defines expected outcomes as well as the organizational boundaries, whereas a micromanager defines and controls the activities that lead to outcomes. There should never be a need to tell any team member how to do their jobs (i.e., micromanage). If so, you have chosen the wrong person for your team. The second mistake is macromanagement or abandonment. Too often leaders develop objectives for their team members and then never follow up. When they eventually check on performance, usually at the end of the year, they are disappointed that many of the objectives were not met.

Active, persistent oversight of progress on objectives is important. It is best to conduct regular periodic meetings individually and in team meetings to determine progress or shortfalls in the performance of objectives and the actions proposed to correct them. Talented managers never confuse motion or activity with progress.

As you will see in later chapters, the performance management process becomes much easier and more successful when you start first with understanding the development objectives of your staff first, making sure that they are continuously aligned with the performance objectives of the organization. When they are aligned, the performance review process becomes self-directed.

As a team leader, you have the responsibility to support the actions proposed and play an active role in overcoming the adversity faced by your team members through continuous problem solving and influence. This is managing from the middle. This includes improving processes and eliminating bottlenecks. In the words of Peter Drucker, such management takes hard work, sweat, and practice. Plans are only good intentions unless they are immediately followed by hard work.

There are a few critical actions that you personally can take to develop early confidence of your team members. Practice what you preach. Make sure that the vast majority of your personal time is spent on the actionable objectives critical for success. This means prioritizing your meeting topics, emails, telephone conversations, and individual meetings to receive the highest priority and most time commitment. A useful exercise is to track your time commitments over a 2-week period and determine how much time was spent on you and your team's actionable objectives. That percentage should be no less than 80%. Your team will follow suit if you lead by example.

Also, achieve short-term wins on some of your own objectives. This will show that you are heading down the right path and that the strategy is initially working. Whenever a milestone or objective is missed (and invariably some will), rather than ignoring it or trying to explain it away, own up to it. It is best to meet with team members and conduct a "lessons-learned" session and develop a revised plan to get back on track. Although there will be some concern about the strategy, your team members will grow more confident in you as a leader who has the courage to acknowledge and learn from your mistakes.

2.7

The single most important reason why the execution of strategies fail is tolerating underperformance

It is important to not tolerate underperformance. *In my opinion, this is the single most important reason why the execution of strategies fails.* If a team member consistently fails to perform despite lessons learned and active support from you, you must remove him from the team no matter how hard it may be. Too many team members are left in their positions contributing little and potentially derailing the team's strategy and demotivating other team members.

In addition to monitoring what team members are accomplishing, it is important to check on how they are achieving their objectives. Evaluating team member's behaviors and whether they engender trust is critical to team performance.

☐ Coaching—Gaining Trust through Development

Many would argue that completing the first two processes of the Performance Trilogy defines success; developing a compelling strategy and successfully executing it. In my experience, however, sustainable leadership requires that you complete the third process, gaining the trust of your team. Leaders can succeed with their initiatives in the short term, despite a lack of integrity and genuine

2.8

Acting with integrity is a critical element in gaining the trust of your team. This involves being honest with your words, consistent in your actions, and keeping your promises.

concern for the success of their team members. Over time, however, this behavior catches up with them. Broken promises are not forgotten. Lack of recognition and career development is resented. As a result, gaining support for new initiatives becomes increasingly more difficult. Leadership cannot be sustained without trust, and trust is gained not only by achieving results with your strategy but also acting with integrity and genuine concern for your subordinates and colleagues [6]. I like to think of this as value-based leadership. Trust is not a "like to have" but a "must have" if you are to achieve extraordinary performance. The rationale is articulated brilliantly in Covey's book *The Speed of Trust* [22].

Acting with integrity is a critical element in gaining the trust of your team. This involves being honest with your words, consistent in your actions, and keeping your promises. During the strategic planning process, you spend time with each of your key team members to determine their aspirations and how successfully executing the strategy may fulfill those aspirations. You developed both extrinsic and intrinsic rewards that fit your team members' expectations and make commitments to them if they performed. Meeting these commitments is as important for sustainable leadership as the success of your strategy.

There will be circumstances, however, when promises may have to be broken and team members are disappointed. This can be due to a deliberate change in plans or assessment of performance. Plans are often modified based on experience, lessons are learned, and sometimes roles are changed. Mentoring staff through this process with honesty and transparency will build trust, provided that you act not from a personal agenda but for what is good for your team and the organization. When tough decisions need to be made and a promise must be broken, spend the time to explain the reasons and take responsibility for your actions. If the decision was based on a shortfall in performance, you need to point this out to the team member and recommend personal development to correct the problem.

In addition to keeping your promises, trust involves a high degree of honesty and consistency in your actions. Never underestimate how closely you are watched and your actions discussed by your team members. Are your words and statements the same from day to day, week to week? Are they the same regardless of which team member or groups you talk to? Are they the same both in public and private? There is an authenticity that you emanate when your words and deeds are consistent over time and place. If you have to change a decision that you have made publicly, you need to spend time and effort explaining why you have changed your mind based on new information and circumstances. This is the antithesis of politicians, who as we all know are not noted for their authenticity and trustworthiness. Authenticity is the key to long-term, sustainable success as described by Bill George in his landmark book *True North* [14].

Last but certainly not least is to demonstrate genuine care for your team. At a fundamental level, we all trust only those people who we think care about us both professionally and personally. Too many leaders treat their team members as objects or "human resources" (a term which I intensely dislike) rather than people and spend way too little time coaching them. When I finally matured enough midway through my career and began to understand that I was only as good as the people who worked for me, the trust level of my team greatly improved, and my effectiveness skyrocketed. I now believe that gaining trust is so important that in my performance workshops and consulting assignments, I give as much time to leadership development that I give to strategy and execution.

At a professional level, it all starts with selecting the right team members. A common mistake made by most technical managers is to select primarily for knowledge and experience. While these factors are important, selecting for talent is even more important. One of the signs of a good leader is the ability to describe, in detail, the specific talents of his or her team members and capitalize on those talents to find the right fit with the available assignments. The more you know about what drives each team member and how each one thinks, the higher the likelihood that you will be able to coach them properly. A leader needs to be a catalyst, turning top talent into performance.

During the execution phase, each team member was given actionable performance objectives that aligned close to 100% with the strategic thrusts. In most organizations, annual performance plans also include development objectives as well as performance objectives. Based on my experience in evaluating development plans, they are less aligned (and in many cases totally unaligned) with strategy than performance objectives. The reason for this I believe is that performance objectives directly affect the leader's ability to successfully execute his or her strategy while the development plans focus on the team member's ability to perform successfully in the future. Caring about a team member's future and coaching him or her to achieve their professional goals goes a long way to building trust. This means mentoring as well as coaching; finding the right fit where the team member will be most productive and happy rather than just climbing the next rung on the organizational ladder.

Finally, on a personal level, understanding a team member's personal circumstances and how they may be affecting his or her professional performance is an important part of showing concern. Too often, leaders make assumptions about performance without taking into account personal circumstances, which can lead to faulty decisions that adversely affect team members. The more aware you are about the reasons behind a person's behavior, the more likely you are to make the right decisions on his or her behalf.

Caring about the well-being of your team members and helping them in achieving high performance will not only build trust but also loyalty making your next leadership assignment that much easier. Current team members will be willing and anxious to follow you to your next assignment and also spread the word that you are the type of leader worth following.

At the beginning of this chapter, I warned you that leading was not easy! It is the main reason why excellent S&T leaders are few and far between. If you are willing to put in the work and not cut corners in executing all three leadership roles of the Performance Trilogy, you will develop a track record as an outstanding leader, increase your sphere of influence, and be ready to tackle more difficult leadership challenges with a higher degree of confidence.

So, whenever I am asked to conduct a performance audit in an S&T organization, I ask four key questions of several layers of management and key staff. The answers tell me a lot about the organization and which of the three essentials to focus on.

1. What is your organization's strategy and tell me whether you believe it will work over time?

2. What are your organization's key institutional objectives (i.e., senior leadership performance goals) for this year in support of the strategy and how confident are you that they will be achieved?

3. What contribution do you plan on making to help achieve the institutional objectives and is it part of your performance plan?

4. How much trust do you have that your manager has your career interests at heart?

The answers to these four questions will pretty much determine how well an organization is practicing the Performance Trilogy and is a pretty good indication of whether it will be successful.

☐ Chapter Summary

Leadership Definition—"Leadership is the art and science of creating the future by inspiring others to follow you to your desired destination, building their confidence in the path taken, and gaining their trust that you have their interests at heart. The essence of leadership is the ability to transform ideas into significant results."

Leadership (capital L) involves understanding and implementing the three roles and responsibilities of the Performance Trilogy.

Leading the strategy (small l)—"lead from the front" to create and promote a compelling strategy that inspires faith among your team. To do this, you must answer the following three questions to the satisfaction of your staff.

Is the destination desirable?—It is not only important to describe "what" the ultimate destination is, but more importantly, "why" it's important and urgent enough that it is worth pursuing.

Is the destination achievable?—Your strategy needs to be well thought out and you and the team have the talent, motivation, skills, and knowledge necessary to successfully execute the strategy.

Is the destination beneficial? You need to spend quality time with your staff to assess what their aspirations are and how the achievement of the strategy will fulfill those aspirations.

Managing the execution—"lead from the middle" to build confidence among your team with your ability to execute the strategy to produce results.

The institutional objectives need to be cascaded down through each management level of the organization.

Manage the execution process by following up with *regular* reviews of your team's actual performance

Be a role model—practice what you preach

Own up to your performance—Admit mistakes and conduct lessons learned.

Never tolerate underperformance

Coaching the development—"lead from the rear" to gain your team's trust by empowering them, guiding them, and helping them grow as leaders.

Leadership cannot be sustained without trust, and trust is gained not only by achieving results with your strategy, but also by acting with integrity and genuine concern for your subordinates and colleagues

Acting with integrity is a critical element in gaining the trust of your team. This involves being honest with your words, consistent in your actions, and keeping your promises

When tough decisions need to be made and a promise must be broken, spend the time to explain the reasons and take responsibility for your actions.

Demonstrate genuine care for your team. At a fundamental level, we all trust only those people who we think care about us both professionally and personally.

☐ References

1. Geneen, Harold with Moscow, Alvin, *Managing*, Doubleday, Garden City, NY, 1984.
2. Hitt, William, *The Model Leader*, Battelle Press, Columbus, Richland, 1993.
3. McCormack, Mark H., *What They Don't Teach You at Harvard Business School: Notes From a Street-smart Executive*, Bantam Books, New York, 1984.
4. Peters, Thomas J., Waterman, Robert H., *In Search of Excellence: Lesson's from America's Best-Run Companies*, Harper & Row Publishers, New York, 1982.
5. Townsend, Robert, *Further Up the Organization: How to Stop Management from Stifling People and Strangling Productivity*. Alfred A. Knopf, New York, 1984.

6. Shaw, Robert Bruce, *Trust in the Balance, Building Successful Organizations on Results, Integrity, and Concern*, Jossey-Bass, San Francisco, CA, 1995.

7. Garfield, Charles, *Peak Performers: The New Heroes of American Business*, William Morrow, New York, 1986.

8. Drucker, Peter, *The Effective Executive*, Harper and Row, New York, 1966. (Updated in 2007 The Effective Executive, The Definitive Guide to Getting the Right Things Done Collins Business)

9. Collins, Jim, *Good to Great*, HarperCollins, New York, 2001.

10. Senge, Peter M., *The Fifth Discipline: The Art & Practice of the Learning Organization*, Doubleday, New York, 1994.

11. Bossidy, Larry, Charan, Ram, *Execution: The Discipline of Getting Things Done*, Crown Business, New York, 2002.

12. Stack, Jack, *The Great Game of Business*, Doubleday, New York, 1992.

13. Labovitz, George, Rosansky, Victor, *The Power of Alignment: How Great Companies Stay Centered and Accomplish Extraordinary Things*, John Wiley, New York, 1997.

14. George, Bill, *Authentic Leadership*, Jossey-Bass, San Francisco, CA, 2003.

15. Goleman, Daniel, *Emotional Intelligence*, Random House, New York, 1995.

16. Bennis, Warren, *On Becoming a Leader*, Perseus, New York, 1989.

17. Buckingham, Marcus, Coffman, Kurt, *First, Break All the Rules, What the World's Greatest Managers Due Differently*, Simon and Shuster, New York, 1999.

18. Katzenbach, Jon R., Smith, Douglas K., *The Wisdom of Teams*, Harvard Business school Press, Boston, MA, 1993.

19. Loehr, Jim, Schwartz, Tony, *The Power of Full Engagement*, Simon and Schuster, New York, 2003.

20. Kotter, John P., *Leading Change*, Harvard Business School Press, Boston, MA, 1996.

21. http://www.ted.com/talks/simon_sinek_how_great_leaders_inspire_action.html.

22. Covey, Stephan M. R., *The Speed of Trust, the One Thing that Changes Everything*, Simon and Shuster, New York, 2006.

Personal Leadership

Personal Mastery through Lifelong
Learning and Self-Discovery

Are You Ready to Lead?
Leadership Is Personal

In Chapter 1 on the Performance Trilogy®, we emphasized focusing on the three fundamental processes of performance: strategy, execution, and leadership development. In Chapter 2, we presented a leadership framework describing the key roles and actions that leaders must take during each process of the Performance Trilogy, including inspiring faith in the strategy, building confidence in the execution, and gaining trust through leadership development.

In this chapter, we're now ready to address the question of "are you ready to lead?" Before you attempt to take on a leadership assignment, whether it is a personal goal, your first team leadership role (e.g., managing a technical project), or advancing to a higher level of management leadership, several important questions need to be asked. While the focus of this discussion is on self-assessment, these same questions can also be used to evaluate team members for future assignments.

On an individual level, most scientists and engineers start out with clear professional leadership goals. These would include working on important and impactful research projects, publishing and presenting their research, achieving recognition among their peers, and moving up the technical ladder to senior technical positions within an organization.

The majority of scientists and engineers I talk to are reluctant to take on a managerial leadership assignment in their organization despite the promotion potential and salary increases. The reasons I hear most often are the loss of status and respect among their technical peers and the distraction from their technical work due to increased administrative duties. Unfortunately, these perceptions are based on experience and reinforced by a genuine lack of good role models. On the other hand, I have seen too many overly ambitious staff seek managerial positions with neither the talent level nor the right motivation just to move one more rung up the organizational ladder to gain power and make more money.

I was fortunate enough early in my career to work for a science and technology (S&T) manager who was a good role model and influenced me to pursue a career in management. To maintain respect from your peers, it is most important to make the transition gradually by excelling at managing projects while still maintaining a technical leadership role in experimental design and data interpretation. An additional step is to take on a group or discipline leader role in addition to your own technical projects where you can begin the transition to management. In this informal "player-coach" role, you can maintain your technical status by providing useful advice on proposed technical initiatives by your team members as well as technical reviews on project reports. At the same time, you can begin to gain valuable experience in leading and coaching staff.

There is no doubt that taking on an S&T managerial position will require an increase in administrative duties, which will increase as you take on larger executive management roles. It is important to keep these duties in perspective. As the ultimate decision maker for your S&T management unit, you will have the responsibility and signature authority to "manage" a large number of organizational processes as described in Chapter 8. Too many managers do get overwhelmed by the amount of information they need to process and the many daily decisions they need to make. They wind up reacting to the pile of requests in their in-basket, emails and text messages rather than having a prioritized system. Where you spend your time and the decisions you make will ultimately determine yours and your organization's success. This will be discussed in more detail in Chapter 6.

S&T management is not only an administrative position but also a critically important leadership function. As an S&T manager, your primary role is to create business and societal value from S&T. With one foot in S&T and the other in management, you serve as a critically important link that provides important leadership input into the development of organizational strategy and translating that strategy into actionable objectives that need to be managed. To achieve those objectives, you must coach your team members who are responsible for getting the work done by providing direction, support, and encouragement. This is the essence of the Performance Trilogy and should consume the vast majority of your time. If you are having trouble focusing and executing the Performance Trilogy in your job because you are overwhelmed by administrative duties, you need to ask for help.

3.1

If you are having trouble focusing and executing the Performance Trilogy in your job because you are overwhelmed by administrative duties, you need to ask for help.

The hardest person you will ever have to lead in your lifetime is yourself. You cannot effectively lead others unless you have begun the lifelong journey of self-discovery that starts with the desire to increase your self-awareness.

The academic literature is filled with theories that attempt to predict leadership success [1–5]. Throughout the past decades, there have been multiple models (personality, traits, contingency, competency, etc.) proposed. None of these models alone have adequately predicted leadership success. Every one of these models has its merits but only addresses a single element of the success equation. Leadership is a complex social phenomenon that is difficult to boil down to simple theories. Sun Tzu said it best when he stated, "Leadership is a matter of intelligence, trustworthiness, humaneness, courage, and discipline... Reliance on intelligence alone results in rebelliousness. Exercise of

3.2

Leadership is a matter of intelligence, trustworthiness, humaneness, courage, and discipline... Reliance on intelligence alone results in rebelliousness. Exercise of humaneness alone results in weakness. Fixation on trust results in folly. Dependence on the strength of courage results in violence. Excessive discipline and sternness in command result in cruelty. When one has all five virtues together, each appropriate to its function, then one can be a leader.

Sun Tzu

humaneness alone results in weakness. Fixation on trust results in folly. Dependence on the strength of courage results in violence. Excessive discipline and sternness in command result in cruelty. When one has all five virtues together, each appropriate to its function, then one can be a leader."

Based on reviewing these models and personal experience, my point of view is that not all opportunities will have an equal potential for success or are right for you at a given point in time. When faced with the decision as to whether to take on a leadership assignment, you need to do your homework and answer two key questions: "What is the magnitude of the challenge presented" and "what key attributes do you bring to the table?"

☐ Estimating the Magnitude of the Challenge

Clearly, not all opportunities or challenges are equal. Developing a clear understanding of the challenge or opportunity you are faced with is the first step in determining whether you are prepared to take on the assignment. Analyzing the magnitude of the challenge can be broken down into three key factors: the degree of difficulty of the assignment, the level of influence you have, and the resources available to you.

Calculate the Degree of Difficulty of the Assignment

The first question that needs to be answered is "how difficult is the assignment?" Oftentimes, this can be addressed by estimating the speed and magnitude of change that is needed. In one extreme, the challenge may be to affect an incremental improvement in performance that is well within your experience and leadership attributes. In the other extreme, the challenge may involve transformational change that requires abandoning familiar practices and taking a whole new approach that involves considerable risks. In between these extremes lies a host of challenges with progressively higher degrees of difficulty (Figure 3.1).

Estimating the Magnitude of Change

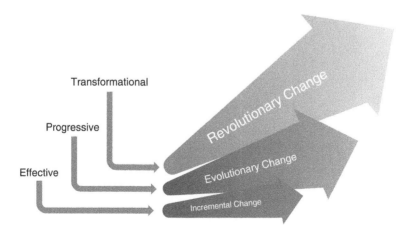

FIGURE 3.1 Estimating the magnitude of the challenge.

I was fortunate enough to have a boss very early in my career that passed me over for an important management assignment that would have been a significant career move for me. While my first reaction was anger, I left his office after 2 hours of dialogue with an appreciation of why he made the decision. He pointed out the specific challenges of the new assignment and that my level of influence and attributes weren't yet robust enough to ensure a reasonable chance of success. He also assured me that my management potential was very high and that it wouldn't be long before the right opportunity came along. While I was disappointed, I had to admit that his rationale made a lot of sense.

Estimate Your Level of Influence

Once you have determined the degree of difficulty of the assignment, the next step is to determine who you will need to help you succeed. Many assignments require only your personal leadership and you will be able to assess whether you have sufficient attributes to succeed. If the assignment involves team leadership, your sphere of influence necessarily broadens to include your peers. Team leadership, leading like-minded professionals, requires a step jump in your level of self-awareness. It is one of the major hurdles faced by technical professionals that assume group management positions. Influence increases with the realization that you do not have to be the best technical professional in the group, only a good "player coach" that can contribute technically but focuses most of your time and leadership skills on the success of the group. Only when you temper your professional ego, will you be able to gain the groups' confidence and trust in your leadership abilities (Figure 3.2).

When moving from team leadership to organizational leadership, your sphere of influence broadens even more to include professions that you know very little about. This represents another step jump in your level of self-awareness. It is another major hurdle

Sphere of Influence

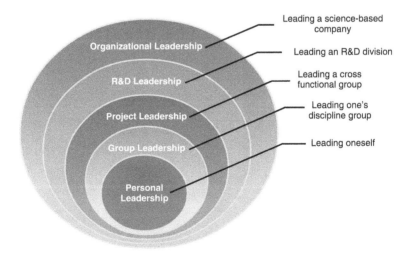

FIGURE 3.2 Estimating your sphere of influence.

faced by technical group managers that assume organizational management positions. Influence increases with the realization that it is not necessary to understand all of the multidisciplinary professions that go into making an organization work. It is more important to take a systems approach to understand the input/output of each organizational unit and manage the interfaces.

In all of the earlier examples, estimating the level of influence of the team that you will be leading is important. Do the team members know you personally? Do they have confidence in you based on your track record? Do they trust you? Have you worked together successfully on previous assignments? The answers to these questions should factor into your decision about your readiness for the assignment.

Inventory Your Assigned Resources

The next step is to estimate the resources that you will need and determine whether they will be available when you need them. Resources include such items as number and qualifications of assigned staff, adequate budgets, physical resources, and commitment from support functions. Commitment is a key element here and you must get a solid commitment from yourself, your manager, and team. Oftentimes, promises are made with good intentions and not delivered, causing you to miss meeting your commitments.

The level of risk of taking on a new assignment is based on the magnitude of the challenge. A high level of risk involves taking on a major change initiative with a tight deadline, very little influence, or trust among team members and limited resources. Succeeding in such a challenge requires a high level of motivation, talent, and skills.

☐ What Are the Key Attributes You Bring to the Leadership Challenge?

In the previous section, the first part of answering the question "are you ready to lead" was to look outward and estimate the magnitude of the challenge. This involves calculating the degree of difficulty of the assignment, estimating your level of influence from those whose help you need, and taking inventory and getting commitment of the necessary resources.

Once the magnitude of the challenge has been determined, you can now look inward to determine whether you have sufficient attributes to meet the leadership challenge. By attributes, I mean the unique qualities and characteristics that define you as a person. From my point of view, the fundamental attributes of high performance are motivation, natural talent, self-awareness, and acquired skills. Based on my experience, most of you will probably focus more on "can I meet the leadership challenge?" by evaluating your past experience and measuring your current skill set. However, the more important question you should ask is "will I meet the challenge?" having much more to do with an understanding of your level of motivation, utilization of your natural talents, and the depth of your self-awareness.

Try not to fall into the "tyranny of how." If you start with asking "how am I going to achieve this goal or succeed in this assignment," you will not have all the answers right away and this will stifle your initiative. On the other hand, if you start with the "what" (i.e., vision and mission) and the "why" (i.e., desirability and benefits), you will more than likely be in a different state of mind and convince yourself that you can figure out the how.

> **3.3**
>
> When determining whether you are ready to lead, start with the what (vision and mission) and why (desirability and benefits) before getting bogged down with the why.

Examine Your Motives

The first key attribute and, in many ways, the most important is motivation. As a young professional, I viewed every promotion opportunity as a way of getting ahead, making more money, and having more prestige. These extrinsic drivers were important to me having grown up in a family with very limited resources. While a healthy dose of ambition can be an initial driver, in my experience is not sufficient to sustain the kind of motivation and extraordinary effort that is needed to ensure successful completion of difficult assignments. After the initial ego gratification and bump in salary wears off, you are left with the day-to-day pressures and difficulties of performing. Peter Drucker's advice is "given the choice between making more money or gaining more experience, always chose gaining more experience."

Later in my career, I became less interested in my personal accomplishments and much more interested in growing high-performance technical organizations that also satisfied the career aspirations of my fellow scientists and engineers. Building teams proved to be much more stimulating and rewarding for me, and as it turns out increases the probability of success. My ambition now is to coach up-and-coming technical managers by passing on the hard lessons I've learned over the years to accelerate their leadership development. This is so rewarding that it doesn't even feel like work anymore.

The cliché "pursue what you love, and you will never have to work a day in your life" has an element of truth in it. Leading is really hard work. Start by asking the question "Why am I interested in this particular assignment besides the money and stature and does its success have meaning and purpose for me?" The answer will help determine whether you have the passion and energy necessary to work at full capacity and persevere through the tough obstacles that you will inevitability encounter.

Discover Your True Talents

There is a strong correlation between what you love to do and your specific talents (i.e., the natural ability to excel at something without being taught). Everyone can become very good at something by working hard at acquiring a skill, but to excel usually requires tapping into your natural talent. At an early age, all of us derived pleasure and satisfaction from the success and positive feedback we got from displaying our natural talents. As a result, we tended to spend more time and attention in nurturing those talents and applying them to our daily life. Some

> **3.4**
>
> Everyone can become very good at something by working hard at acquiring a skill, but to excel usually requires tapping into your natural talent.

realize at an early age that they are more likely to be happier and probably more success-ful if they choose careers that require the use of their specific talents. Their choice may or may not have anything to do with how much money they can make. Others, like myself, focused at first on the economic return that specific careers could offer and had to read-just when the money just wasn't enough to motivate us to achieve at our full potential.

I discovered early on that two of my more obvious talents were my curiosity and skill in mathematics. These talents (critical thinking, problem solving, analytical think-ing, quantitative reasoning) along with my desire for getting ahead and making money drove me toward a career in science and engineering (chemical engineers at that time topped the annual professional earnings list). While I was pretty successful early in my career as a bench scientist, I found that it wasn't as fulfilling as I would have liked. While it satisfied my love for exploration and discovery, the progress of science is very slow and the day-to-day activity required little interaction with others. Given my gregarious nature and lack of patience, I knew that I had made a career mistake. With additional introspection, and experience, I discovered that, with my scientific base, I could take advantage of my other talents. My competitiveness and salesmanship talents allowed me to move out of the laboratory and into business development, which I loved. Writing a complex proposal for funding with my colleagues and winning competitive contracts was almost as much fun as the sandlot touch football games I enjoyed so much as a teenager.

I now have a greater appreciation and a great deal of respect for individuals who persist in following their dreams regardless of the level of remuneration. In my book, they will achieve far greater rewards than a bigger car or a larger house.

Discovering your true talent is not an exercise to be taken lightly. It took me years of analysis and practice to increase my self-awareness before I found out what my real "ichigai" was. Ichigai is an Okinawan expression for the reason why you get out of bed every morning. I receive the greatest amount of pleasure and satisfaction from two activi-ties: exploration and teaching.

Talent comes in a variety of different physical and mental forms if you look hard enough. It can be physical (an imposing stature, an eloquent voice, and physical strength and endurance); mental (a photographic memory, meticulousness, perceptiveness, ease with language, pattern recognition, reading comprehension, mathematics); emotional (patience, empathy, resilience, humor, courage, persistence); and spiritual (intuition, clairvoyance, sense of purpose). Spend time in self-discovery to find your unique set of talents and determine how they can be best used to achieve your goals.

The Importance of Self-Awareness

We have discussed the importance of meeting the leadership challenge by first answering the question, "Will I lead?" dealing with the attributes of motivation and talent before answering the question, "Can I lead?" which has more to do with your acquired skills and experience. The answers you come up with to both of those questions are highly depen-dent on your degree of self-awareness. Before deciding on the direction your career should take, spend considerable time in finding out who you really are. Self-awareness plays an essential role in determining intrinsic motivation, building a healthy self-confidence, and increasing self-esteem. The more you truly know yourself, the better your life decisions will become. I will only scratch the surface of this extremely important, complex, and underappreciated topic, but hopefully will stimulate you to explore further [6,7].

While there has never been a strong correlation between personality traits and leadership success, I believe that a universal attribute of successful leaders is a healthy degree of self-confidence and genuine self-esteem. By *healthy* self-confidence, I mean a leader who knows what he knows and knows what he doesn't know and is not afraid to make prudent decisions based on his current level of awareness. This is opposed to an arrogant leader whose limited awareness permits his confidence level to exceed his abilities and is unable to admit that there are others who know more than he does. This is commonly a result of early successes that go to his head.

> **3.5**
>
> By *healthy* self-confidence, I mean a leader who knows what he knows and knows what he doesn't know and is not afraid to make prudent decisions based on his current level of awareness.

By *genuine* self-esteem, I mean a sense of comfort in one's own skin and motivated by one's own internal values rather than extrinsic rewards and the admiration of others. Bill George in his book, *True North* [8], describes this person as an authentic leader. True leaders, regardless of their personalities, exude self-confidence and self-esteem that make others trust them and want to follow them in their endeavors. Steven Covey describes this well in his book, *The Speed of Trust* [9].

Why is self-awareness so important? The higher one's self-awareness is, the better decisions one makes in life. The cumulative positive effects of good decisions build self-confidence. The higher one's self-awareness, the easier it is to gain favorable reactions from people. Gaining favorable reactions from people helps builds one's self-esteem.

What Is Self-Awareness?

Social psychologists define self-awareness as the degree of clarity with which we perceive, understand, and evaluate, both consciously and nonconsciously everything that affects our lives. While the definition is accurate, it took me years to appreciate the depth of its meaning. Throughout history, sages have expounded the importance of increasing self-awareness through self-discovery and lifelong learning. Quotes such as "know thyself," to "thine own self be true," and the "unexamined life is not worth living" are familiar to all of us.

Unfortunately, cognitive knowledge doesn't translate well to experiential knowledge. It is extremely difficult to see ourselves as others see us, and this is essential to increase our self-awareness. It is said that an hour spent in dialogue with a wise man is worth thousands of hours reading books.

> **3.6**
>
> An hour spent in dialog with a wise man is worth thousands of hours reading books.

I like to think of self-awareness in three parts: the intellectual and mental "database" that we have in our head, the emotional and spiritual "database" that we carry in our hearts, and the filters we use to selectively screen and store information in these two databases. All of us have very thick filters conditioned by our heredity, upbringing, intuitional insights, and total life experience that limits our self-awareness of reality. This limited awareness at times results in faulty decisions and inappropriate behavior. Many of us trained in science and engineering tend to be data rich in our intellectual and mental folders and data poor in our emotional and spiritual folders. As a result, we tend to make most decisions based primarily on "logic," with little regard to intuition and rarely consider the effect of our behavior on others. There is an emerging school of thought that characterizes intuition as the intelligence of the heart [10].

Why Is Self-Awareness So Important?

Bill George reports that 75 members of the Stanford Graduate School of Business Advisory Council nearly unanimously chose self-awareness as the most important capability for leaders to develop [11]. In his book, *Emotional Intelligence*, Daniel Coleman lists self-awareness as the first element in describing the leader's role [12]. Successful leaders do two things exceptionally well. They make good decisions most of the time and build strong and lasting relationships. Increased self-awareness almost always leads to better decisions that in turn reinforce one's self-confidence. Increased self-confidence leads to building a healthy self-esteem that leads to stronger personal relationships. Covey describes the importance of becoming independent before seeking interdependence in his book, *The Seven Habits of Highly Effective People* [13].

The lack of awareness of the factors that influence our behavior oftentimes leads to unfavorable reactions from people. It is also the major cause of faulty decision making. Being aware of and learning from the consequences of your decisions and actions lead to wisdom.

3.7

Being aware of and learning from the consequences of your decisions and actions lead to wisdom.

I was first exposed to the importance of the concept of self-awareness in my 30s while taking a course on Building Self-Esteem from the Barksdale Foundation [14]. It was then that I discovered the uncomfortable truth that I was succeeding in spite of my behavior, not because of it. My actions were designed to prove to the world how smart I was and my decisions at the time were based solely on completing the task at hand and advancing my career with little regard for how my behavior was affecting others around me. By opening up to people who knew me well, I became painfully aware that I was branded as a "know-it-all" and a person that couldn't be trusted. I was fortunate to latch on to a mentor whom I worked with for the next 20 years, who steered me on a path of lifelong learning and discovery. I am still on that path knowing that the journey lasts forever.

Increasing Your Self-Awareness through Exploration and Lifelong Learning

Our awareness is made up of four key factors:

1. Our intellectual acumen: the ability to analyze, correlate, and evaluate all experiences and to accurately determine the cost and benefit of any action we may take.

2. Our intuitional insights driven primarily from subconscious drives and urges emanating from the heart.

3. Our total life conditioning, both conscious and nonconscious, resulting in our knowledge base, values, ideals, and belief system.

4. Our sense of personal worth that drives our moods, attitudes, emotional reactions, prejudices, habits, desires, fears, aspirations, and goals as a result of our conditioning.

The hardest person you will ever have to lead is yourself. You must take personal responsibility for your own development starting with increasing your self-awareness.

Save your money and avoid all advice from instant success books and seminars. Expect the development process to last a lifetime and enjoy the exploration. Some recommendations I can make based on personal experience are the following:

- Become more introspective—find out what really motivates you, what drives your behavior, why do you do the things you do. In time, you will discover your hidden strengths and become aware of your not so obvious weaknesses.

- Find a good mentor—as I said previously, an hour spent with a wise man is worth a thousand hours reading books. One of the hardest things to do is to see ourselves as others see us. A good mentor should make you feel uncomfortable at times.

- Observe and model leadership behavior—watch what outstanding leaders do in a variety of situations and read autobiographies.

- Be willing to dent your ego—listen to feedback from others even if you don't want to hear it. When you experience unfavorable reaction from people, instead of rationalizing, ask yourself "what was it that I said or did that triggered that reaction."

- Seek continuous improvement in your self-awareness—constantly evaluate your decisions and actions by conducting a "lessons-learned" exercise. Decide what went right and what went wrong and objectively examine the reason why. Then commit to improving that decision or action the next time. This is the process of continuous improvement and lifelong learning.

Another variation of the Performance Trilogy is shown in Figure 3.3. When you plan, you strategize; when you do, you execute; and when you learn, you develop, which continuously increases your self-awareness (Figure 3.3).

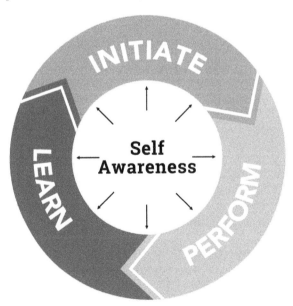

FIGURE 3.3 Personal leadership development.

On my wall in my office is a plaque with my favorite reminder, "better than yesterday, not as good as tomorrow." Once you start on this lifelong journey of self-discovery, I assure you, you will find it personally rewarding.

☐ Chapter Summary

R&D management is not just an administrative position but also a critically important leadership function. As an R&D manager, your primary role is to create business and societal value from S&T.

When faced with the decision as to whether to take on a leadership assignment, you need to do your homework and answer two key questions. "What is the magnitude of the challenge presented" and "what key attributes do you bring to the table?"

What is the magnitude of the challenge?

Calculate the degree of difficulty of the assignment

Estimate your level of influence

Inventory your assigned resources

What are the key attributes you bring to the leadership challenge?

Examine your motives

Discover your true talents

Increase your level of awareness

Self-awareness comes in three parts: the intellectual and mental "database" that we have in our head, the emotional and spiritual "database" that we carry in our hearts, and the filters we use to selectively screen and store information in these two databases.

☐ References

1. Zaccaro, Stephen J., Kemp, Carry, Bader, Paige. Leader traits and attributes. In *The Nature of Leadership*, (pp. 101–124). Sage Publications, Thousand Oaks, CA, 2004.
2. Yukl, Gary A., Van Fleet, David D. Theory and research on leadership in organizations. In *Handbook of Industrial and Organizational Psychology*, Vol. 3 (2nd ed.) (pp. 147–197). Consulting Psychologists Press, Sunnyvale, CA, 1992.
3. Judge, Timothy A., Bono, Joyce E., Ilies, Remus, Gerhardt, Megan W. Personality and leadership: A qualitative and quantitative review. *Journal of Applied Psychology*, 87.4 (2002), 765–780.
4. Blake, Robert R., Mouton, Jane S., *The Managerial Grid: The Key to Leadership Excellence*, Gulf Publishing, Houston, TX, 1964.
5. Bass, Bernard M., Bass, Ruth, *The Bass Handbook of Leadership: Theory, Research, and Managerial Applications* (4th ed.), Free Press, New York, 2008.

6. Duval, Shelley, Wicklund, Robert A., *A Theory of Objective Self-Awareness*, Academic Press, Waltham, MA, 1972.

7. Goleman, Daniel, *Emotional Intelligence,* Bantam Dell, Division of Random House, New York, 1995.

8. George, Bill, *Discover Your True North*, Wiley and Sons, Hoboken, NJ, 2015.

9. Covey, Stephan M.R., *The Speed of Trust, the One Thing That Changes Everything*, Simon and Shuster, New York, 2006.

10. Childre, Doc, Martin, Howard, *The Heart Math Solution*, HarperCollins, San Fransisco, CA, 2011.

11. George, Bill, *Authentic Leadership,* Jossey-Bass, San Francisco, CA, 2003.

12. Goleman, Daniel, *Emotional Intelligence,* Random House, New York, 1995.

13. Covey, Stephen R., *The Seven Habits of Highly Effective People,* Fireside (Simon and Shuster), New York, 1989.

14. Barksdale, Lilburn S., Building self esteem, The Barksdale Foundation, 1989 (can be obtained from the Lifemanagementalliance.com).

CHAPTER

Leadership Attributes
Nine Essential Attributes That Ensure Success

Leading each of the three phases of the Performance Trilogy® requires different attributes (i.e., talents and skills ascribed to you) during each phase. In the leadership phase, your role is to "lead from the front," to create and promote a compelling strategy that inspires faith among your team. *Leaders change things*. They anticipate and proactively set a new path. This is the traditional view of "leadership."

In the management phase, your role is to "lead from the middle" to build confidence among your team with your ability to execute the strategy to produce results. *Managers get things done*. This is the traditional view of "management." In the coaching phase, your role is to "lead from the rear" to gain your team's trust by empowering them, guiding them, and helping them grow as leaders. *Coaches develop people*. This is the traditional view of "coaching." Each of these three phases of leadership requires distinctly different attributes to ensure success (Figure 4.1).

4.1

Leaders change things
Managers get things done
Coaches develop people

☐ Critical Leadership Attributes

In Chapter 2, a Leadership Framework was introduced providing a comprehensive view of the actions a leader must take to ensure faith in his proposed strategy. *By faith, I mean believing in something that you cannot prove*. In this chapter, we will discuss the critical attributes needed to ensure that the appropriate actions are taken. Only when your team has faith in your strategy and a "shared vision" of the destination can you confidently proceed to the execution phase. Changing the status quo as we all know is extremely difficult and requires *imagination* to create a new future different from the present, *courage* to challenge accepted practices despite strong resistance, and *persuasiveness* and strong communication skills to attract a legion of followers.

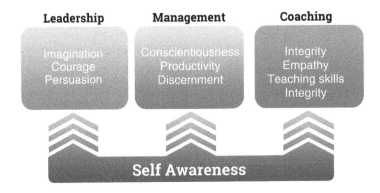

FIGURE 4.1 The nine leadership attributes.

Imagination—The Ability to Look at What Everyone Else Looks at, and to See What No One Else Sees

In Chapter 1, leadership was defined as the *art and science of creating the future by inspiring others to follow you to your desired destination, building their confidence in the path taken, and gaining their trust that you have their interest at heart. The essence of leadership is the ability to transform ideas into significant results.*

The key word for purposes of this discussion is *ideas.* The critical attribute that underpins all others when it comes to developing a strategy is to have a compelling idea that will motivate you to take action. Ideas can emanate from lots of sources but clearly require a vision of a future state and a strategy on how to get there. This strategy will oftentimes require significant changes in direction: requiring new processes and different behavior (hence the often-used business term, change management).

Creating the future is the process of *imagining* what it looks like, determining why it is different than today, and developing a strategy to get you there. There are academic treatises on simulation theory [1], but what it boils down to is your ability to look at what everyone looks at and see what no one else sees. This requires imagination, creativity, and oftentimes inspiration. There can be no doubt that this is a talent (an innate skill) that some are blessed with. It is this attribute that is most often cited by the proponents of the "leaders are born and not made" theory. I don't totally subscribe to this theory and believe that creativity, in addition to being a natural talent, is also an acquired skill that can be improved upon with practice. Sometimes, ideas are inspired as a result of a life experience. My ideas are a lot better today than they were 10 years ago. While not everyone will be able to come up with brilliant ideas, most can generate good ideas that can result in a compelling vision and strategy.

> 4.2
>
> Strategic thinking starts when you spend lots of time in the future imaging things that don't yet exist.

Strategic thinking starts when you spend lots of time in the future imagining things that don't yet exist. In my case, I spend a considerable time in the future, thinking a lot about how to solve problems or identify situations could be made better and how one would go about accomplishing them. Sometimes it's a result of a science article I have read or a business opportunity that looks promising. I must admit that I probably spend a little too much time in the future at the expense of the present. My family often accuses me of being the "absent-minded professor" because my mind is always somewhere else when I should be present with them. My wife is helping me spend more time in the present through meditation and focusing on developing an awareness of my surroundings. God bless her.

For most, however, very little time is spent in the future. This is especially true in developing countries. One of my more colorful Latin American clients was fond of saying that "planning is an exotic word in my culture." Also, many first-time supervisors and middle managers have been trained to focus on present operational issues and develop a healthy disdain for strategic planners (or "dreamers" using their vocabulary). The trick is to know which phase of the Performance Trilogy you are in and focus on those attributes most needed at the time.

In the previous chapter, I pointed out the importance of increasing one's self-awareness. This is an excellent way to develop your critical thinking skills that lead to useful ideas. Force yourself into the future by asking questions. "Why is this not working? How can I improve it? What do I need to know to understand it better? Who can I discuss this with that would give me some good ideas?" In their book, *Creating Excellence,* Hickman and Sliva point out that asking questions that others may not think of asking can lead to creative insight [2]. Such insight forces you to move away from reliance on rules, logic, and efficiency. Listen to different voices, ask different questions, try new experiments, and gain new perspectives. It enables you to get at the heart of a problem, discover hidden opportunities, and opens the door to the best strategies.

While not everyone has the talent to change the world, most of us have sufficient talent along with acquired skills to make the world a little better.

Courage—The Ability to Act Despite Considerable Risks

While ideas are the lifeblood of leadership, the ability to turn those ideas into results takes courage. New ideas are threatening. As Machiavelli [3] is most often quoted as saying "There is nothing more difficult to take in hand, more perilous to conduct, or more uncertain in its success, than to take the lead in the introduction of a new order of things." How often have you heard reactions to

4.3

Ideas are like potential energy. In and of themselves, they are powerful. But only when potential energy is converted to kinetic energy can any useful work gets done.

ideas such as "that will never work," "we've already tried that before and it failed," and "we don't have the experience." As a result, most leaders are constantly faced with the fear of failure. When confronted with this negativity, most leaders recognize that these are different times in a newer world with different people.

I've mentioned that I spend lots of time thinking of new ideas. Early in my career, however, I was extremely cautious and rarely acted on my ideas for fear of failure. The turning point for me was Tony Robbins' concept of the power of taking action in his two books *Unlimited Power* and *Awaken the Giant Within* [4,5]. After reading his books and listening to his tapes, my awareness of the importance of taking action was significantly increased. It dawned on me that ideas are like potential energy. In and of themselves, they are powerful. But only when potential energy is converted to kinetic energy can any useful work get done. Likewise, only when ideas are acted upon, can leaders create change.

I also resonated with Robbins' comment that there is no such thing as failure, only outcomes. Some outcomes are desirable, and you move on, others are undesirable, and you learn from them. So, the only two outcomes result in accomplishment or learning. The more I thought about it, the more I realized what a sound concept that is to overcome the fear of failure. As an experimental scientist, I never worried about a blow to my self-esteem when one of my experiments didn't validate my hypothesis; I just learned from it and tried a new experiment. I now treat every initiative or decision that I make in life as an experiment. I determine (not judge) its outcome and learn from it.

Another important principle that I learned from Robbins was the tyranny of how. When considering a new idea or initiative, never start with the word "how." This is the single biggest reason why most ideas or initiatives never get off the ground. When you start with "what" you want to accomplish (vision) and "why" you must accomplish it (motivation), you will invariably figure out the "how."

> **4.4**
>
> When you start with "what" you want to accomplish (vision) and "why" you must accomplish it (motivation), you will invariably figure out the "how."

Lest I be accused of being too idealistic, I refer to courage in the Aristotelian sense "the habit of choosing the mean between the two extremes" [6]. On one extreme is timidity or the lack of courage. When the challenge is hard and the resistance is strong, courage is an essential attribute of leadership. The other extreme is foolhardiness or taking risks without first understanding your capabilities and resources and also assessing the consequences. Courage in the Aristotelian sense is taking calculated risks.

Persuasiveness—The Ability to Influence

When soliciting venture capital financing for a new company, we were spinning out of Arthur D. Little, my awareness was greatly increased of the power of persuasive communication. We spent a considerable amount of time preparing for these "dog and pony shows" by sharpening our message about the value of our technology and the robustness of our business plan. We were enlightened by one reviewer who told us, "Look, ideas are a dime a dozen, and most of them are pretty good; we make investment decisions based on people and not ideas. I'm a busy man. You have five more minutes to convince me that I should invest in YOU!" The lesson we learned that day can be summed up in one of Steven Covey's seven habits of successful people, "seek to understand, before seeking to be understood" [7].

In Chapter 2, we discussed the importance of inspiring others to have faith in your ideas. To successfully persuade others to have faith in your ideas, you must have a strategy that is desirable, achievable, and beneficial in their eyes. People do things for their reasons, not yours. Hickman and Silva in *Creating Excellence* [8] point out that people are not all the same, and they're certainly not the same as you. To persuade someone to adopt your ideas, you must look inside each individual to gain knowledge of that person's expectations and needs to understand how best to convey your message. I am reminded of a colleague who complained that "our sector leader claims to have communicated his new strategy to us, but all I got

> **4.5**
>
> Our sector leader claims to have communicated his new strategy to us, but all I got was a memo!

was a memo!" Leaders that communicate well understand the four steps of the transmission and reception process. The four steps are

1. Meaning—What is it that I want to say? Is it clear and unambiguous? Do I mean it?

2. Speaking—How am I translating what I mean in both words and body language?

3. Reception—What did the other person hear me say? How might he/she have filtered it based on his or her current awareness?

4. Understanding—How did the other person interpret what he heard?

How often have you overheard the following reactions to conversations? "Didn't you hear what I just said!" "That's not what I meant!" "What are you talking about!" These miscommunications arise when you focus only on transmission and pay little attention to reception. I repeat Covey's quote: "seek to understand before seeking to be understood." This is best accomplished by a technique that most of us know but rarely practice called "active listening" (Figure 4.2).

Listening is hard work intellectually and emotionally and is more tiring than talking because it requires total concentration. I must admit that listening is one of my serious weaknesses. While people are speaking to me at approximately 100 words a minute, I oftentimes hear what they say, interpret what they mean, and interrupt before they even finish what they were going to say. On those occasions, when I do practice active listening, my ability to communicate dramatically improves and the results I get are significantly better. It's a work in progress for me.

Upon finishing your message (transmission), ask the person or group to whom you are speaking to tell you what they heard (reception). This is an effective way to probe not only what message they heard but also how they interpreted it in their own minds. When listening to someone speak, paraphrase what they said to provide feedback on your interpretation of their meaning. Sometimes it will take two or three iterations to achieve true communication. You know you are successfully communicating when your audience tells you "We heard what you said and understand what you meant." This is the true meaning of the expression "meeting of the minds."

Communication is a Four Step Process

FIGURE 4.2 Communication is a four-step process.

☐ Critical Management Attributes

In the management phase of the Performance Trilogy, the manager's role is to "lead from the middle" to build confidence among his/her team with the ability to execute the strategy developed to produce results. In many ways, the attributes required for management are at opposite poles from the leadership attributes. While the leader's mission is to change things, the manager's mission is to standardize and optimize to make things work.

In this section, we will discuss the critical attributes needed to ensure that the appropriate management actions are taken. Only when your team develops confidence in your ability to execute can a strategy be successfully implemented. Producing results requires *conscientiousness*, the motivation and discipline to be thorough and dependable; *productivity*, the ability to manage one's energy and focus; and *discernment*, the ability to consistently make good decisions based on cognitive and intuitive judgment.

Conscientiousness—The Motivation and Discipline to Be Thorough and Dependable

Woody Allen was famously quoted as saying that "80% of success is just showing up." While on the surface this seems like a joke, I believe there is a lot of hidden meaning to this expression. A critical attribute of excellent managers is their discipline and consistency in "showing up." The role of managers is to control processes and projects (as opposed to leaders whose role it is to create change and coaches who develop people). Once a new change process is created, the manager's job is to control all the variables to make that process work to produce results.

4.6

80% of success is just showing up.

Conscientious managers plan ahead, delegate responsibly, pay attention to detail, and follow up in a disciplined way to ensure that processes and projects are proceeding according to plan. In Chapter 10, we will be describing the manager's role in the performance management process and the five critical steps to ensure that the strategy is being properly executed. It involves translating strategy into an execution plan with specific performance objectives and monitoring those objectives in a disciplined way.

Conscientious managers understand the importance of rigorously following policies and procedures and establish key performance indicators to ensure consistent quality and uniformity across the organization. A common mantra of these managers is "what gets measured, gets managed" [10]. This kind of attention to detail is universally present in conscientious managers. I always looked for this attention to detail in my technical project managers, delivering high quality reports on time and on budget. This indicated to me that the person had the potential to become a technical group leader and eventually a higher-level manager.

Punctuality and preparedness are additional traits of conscientious managers. They show a great deal of respect for their own and others' valuable time and always seem to be prepared (see the next section on productivity). I will admit to being fairly undisciplined earlier in my career and as I reflect on it now, it was probably the reason for being passed over for promotion on a few occasions. Recognizing this weakness, I gravitated towards turnaround assignments that required leadership attributes, new transformational strategies, and dramatic change management actions. Once the organization was back on its feet, I passed it on to more a disciplined manager.

There is a certain irony in the fact that it is the very qualities of a conscientious manager that oftentimes prevents him or her from properly executing a newly developed strategy that requires change. It is uncomfortable to have to change a system or process that is working for the sake of a new untried initiative. After all, products and services still need to be delivered despite the disruptive changes being requested. I often describe the process as trying to change a flat tire while the car is still moving!

Productivity—The Ability to Manage One's Energy and Focus

Newly appointed managers often feel overwhelmed by the amount of responsibility they have acquired and the work needed to be done to stay on top of things. Technical staff accustomed to focusing on a selectively few subjects suddenly find themselves overwhelmed daily with tens to hundreds of "events." Making a successful transition from technical expert to manager requires awareness that there are not enough hours in the day to play an active role in every item that comes across your desk. The quicker that one learns how to prioritize and delegate (i.e., get organized) the faster they will absorb the responsibilities of the job and execute productively.

Newly appointed managers find out quickly that the scarcest resource they have to manage is their own time. Being productive is often equated to managing your time. The term "time management" is a misnomer as we all know that time cannot be managed. The minutes and hours of every day will pass according to schedule regardless of our actions. A better way to view being productive is to effectively and efficiently manage the events (or tasks) that we are involved in and the information we are required to process. In a typical day, most managers are whipsawed from one event to another without time to spend on any one of them. In addition, they are responsible for processing an enormous amount of information that comes across their desk or email. Fortunately, methods and tools have been developed to help managers cope with the volume of events and information to be processed and acted upon. If you are still using the same methods and tools to manage events and information that you used 10 years ago, I assure you that you will not be able to keep up with the speed of your colleagues that have acquired the latest productivity tools.

The short overview of the latest organizational tools involves two critical steps: effective prioritizing and efficient processing. In his book, *The Seven Habits of Successful People*, Steven Covey provides an excellent tool for prioritization taken from the Eisenhower Matrix [7]. He characterizes all events and information in a two-dimensional matrix: those that are important and those that are urgent. The first priority is to address all those events and information that are both urgent and important. These are events that you must tackle or information you must process immediately. It is up to you to determine the importance of any event or piece of information in all aspects of your life not just work. I consistently advise managers to list the top ten priorities in their life and to determine how much time they allocate to each one on a weekly basis. The results of this exercise are revealing and often shocking. In their professional work environment, I insist that they make sure that their annual performance objectives are close to 100% aligned with their strategy and they allocate a majority of every workday on advancing these objectives. While this seems intuitively obvious, I have rarely seen this practiced (Figure 4.3).

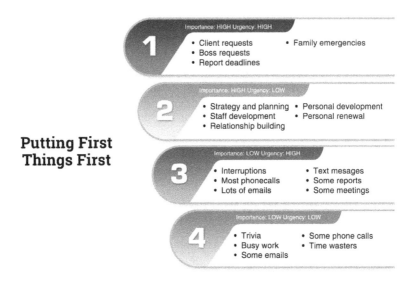

Putting First Things First

FIGURE 4.3 The Eisenhower matrix.

The second priority is those events that are important but not urgent. These events are the most neglected and in many ways the most important for effective prioritization. If you reflect on this category, it represents those events and information that are important but do not have to be done right away. In time however, they will eventually move into the important and urgent category. The more that you can tackle important events and information before they become urgent, the more on top of your job you will be and the less stress you will cause yourself and those who need to support you. This requires a planning discipline and the ability to avoid procrastination. For example, most managers list family as a highly important priority, but when evaluating how much time they spend with their family on a weekly basis, they fall way short of expectations. The excuse that they have too much work to do is a poor one when one adds up all of the time they spend on nonimportant and nonurgent activities that have little to do with important work activities (a quick check on the time spent on answering unimportant phone calls and emails will quickly prove my point).

The third and fourth priorities, urgent but not important and not urgent and not important, are the categories that are the major time wasters and, once identified, can really free up one's time. Lest you interpret the earlier exercise as leading to all work and no play, let me offer you a different point of view to think about, "only the organized can loaf." By only focusing on the important (including recovery and relaxation time) and urgent priorities, you will experience less stress and more freedom to pursue those activities that really count.

Once you have completed step one and prioritized your workload, you can then focus on step two, processing that workload in an efficient manner. Bear in mind that prioritization is a never-ending process and you must reprioritize regularly. Even when you have identified all the important events that you need to pay attention to (both urgent and nonurgent), there is still an enormous amount of work to be done. Despite how good a memory you may have, when you get to a managerial level, the number of events you need to stay on top of and the amount of information you need to process is still daunting. You have probably figured out those traditional tools such as to-do lists and calendars are no longer sufficient to manage all the important events and information in your life. I highly recommend reading David Allen's book on *Getting Things Done* [10]. It is a systematic, no nonsense approach to efficiently acquiring, processing, and storing information, and retrieving it when you need it.

A sure sign that a person is overwhelmed with his job and not utilizing the latest tools is when you get a request to resend an email, to extend the date of a deliverable, or an apology for being a bottleneck in reviewing a report. The problem is not a lack of time but a lack of clarity on the actions needed to produce the desired outcomes. Allen's approach is twofold: capture all of the information and events that you need to manage in a trusted system outside your mind so that you know you can come back to regularly, kind of like an external hard drive, and disciplining yourself to make front-end decisions about all of the inputs that you let into your life so that you will always have a plan for the next actions that you can implement at any moment.

It is a waste of time and highly stressful to be thinking about something that you make no progress on and fearing that you have forgotten something important. When you have a trusted system that stores all of your important information with the ability to retrieve it whenever you wish, you then can free up your mind (or your available RAM) to have the ability to dedicate 100% of your attention to events or information of your choosing with no distraction. This was a giant leap in my level of awareness and helped me tremendously in becoming more organized and productive, getting a lot more done with less effort. It is quite liberating to be able to take on new commitments, knowing that you have the time available based on reviewing your entire workload.

This subject is so important, I will add to the discussion in Chapter 6, the Art of Supervision.

Discernment—The Ability to Consistently Make Good Decisions Based on Cognitive and Intuitive Judgment

Thomas Huxley is quoted as saying "science is simply common sense at its best, that is, rigidly accurate in observation, and merciless to fallacy in logic." The irony of this quote is that scientists and engineers oftentimes are enamored with expert opinions and objective data while ignoring the intuitively obvious. This has led to a more famous

> **4.7**
>
> Discernment is the ability to choose the best possible alternative after weighing both the positives and negatives of a decision.

quote by Voltaire that "common sense is not so common," pointing out the lack of discernment that goes into most decisions. Despite the best-laid plans, invariably things never seem to go according to plan. When executing, a critical success factor for managers is to make a high percentage of good decisions. It is the collective decisions made each day that determine the success of an organization over time.

Discernment is the ability to choose the best possible alternative after weighing both the positives and negatives of a decision. It is judgment based on facts and impressions based on your self-awareness. More often than not, managers are faced with choosing between the lesser of two evils. As scientists and engineers, we are trained not to develop conclusions without first obtaining all of the data. While this rule is essential in a technical context, managers rarely have the luxury of having all the information they would like when making decisions. Too often, technical managers have difficulty making decisions with limited data and procrastinate. This rarely leads to better decision making. In his book *Blink* [11], Gladwell makes a convincing argument that the first decision that comes into one's mind is most often the best one: that instinctual thinking is not capricious but based on the brain's ability to process both consciously and unconsciously the sum total of all your experience.

My decision making improved dramatically when I considered both the immediate outcome of a particular decision and the longer-term consequences of the decision on the people involved. Early in my career, the most important decision criteria for me were to ensure an immediate success of the task at hand, regardless of the longer-term consequences. Over time, I developed an attitude that when making a tough decision, more often than not, it is better to be kind than to be right. While this may seem counterintuitive at first, I found that I have been much more satisfied with the results.

As an example, I had to approve a very expensive decision to bid a large government contract made by one of my managers. I was convinced that we didn't have much of a chance of winning and that the labor cost we would incur would be detrimental to our bottom line. After much debate, I approved the bid decision. My rationale was that the manager was so enthusiastic about the contract and so motivated to write a good proposal that the long-term consequences of turning him down and demotivating one of my more talented managers outweighed the short-term hit to our bottom line. You will never find this kind of decision making in any course on managing proposals. Much to my surprise, we actually won the contract, but even if we had not, the decision was the correct one in my mind.

☐ Critical Coaching Attributes

In this section, we will complete the discussion on leadership attributes with the critical attributes of coaching. Phase 3 of the Performance Trilogy coaches "lead from behind" by developing the team during the execution of the strategy and building the trust of team members. Building trust requires *integrity*, behavior that is honest and consistent in words and actions; *empathy*, a genuine concern for the feelings and aspirations of others; and *teaching skills*, the desire to learn the talents and motivations of others and stimulate their self-learning.

4.8

Before you become a leader, success is all about growing yourself. When you become a leader, success is all about growing others

The secret to sustainable leadership is gaining the trust of your team. When leading the strategy and managing the execution, it is all about you and achieving your goals. When coaching development, it is no longer about you but the success and development of your team. In the words of Jack Welch, "before you become a leader, success is all about growing yourself. When you become a leader, success is all about growing others" [12]. It is possible for leaders to succeed in the short term, despite a lack of integrity and genuine concern for the success of their team members. Over time, however, this behavior catches up with them. Broken promises are not forgotten. Team members resent lack of recognition and career development. As a result, gaining support for new initiatives becomes increasingly more difficult.

4.9

When managing, you lead from the head to advance your personal agenda. When coaching, you lead the heart to support the agenda of others.

Leadership cannot be sustained without trust, and trust is gained not only by achieving your personal goals but also by acting with integrity and genuine concern for the success and development of your subordinates and colleagues. When leading and managing, you "lead from the head" to advance your personal agenda. When coaching, you "lead from the heart" to support the agendas of others. In his excellent book *True North*, Bill George calls this "values-based leadership" [13].

Integrity—Behavior That Is Genuine in Words and Actions

Integrity is the foundation upon which all trust is based. We may respect and admire successful leaders but only trust those leaders who are honest in what they say and consistent in how they act. While no one that I am aware of is a saint, there are several leaders who I had the fortune of working for embraced a set of values and were very consistent in practicing those values. You always knew where they stood on most issues and could depend on their consistency. I may not have always agreed with their decisions but always respected the fact that they were true to their business ethics. An excellent source on the role of integrity in building trust can be found in *Trust in the Balance* by Robert Bruce Shaw [14].

Consistency is an important element in building trust. When we observe inconsistency in what someone says or how they behave over time and different situations, we tend to question whether their motives are pure. I have always espoused the principle that "one should never attribute malevolence to anyone when their actions can be more readily explained by incompetence or inattention." Unfortunately, it is human nature to mistrust those who do not act consistently. "Walking the talk" should not be just a slogan, but a way of life.

4.10

One should never attribute malevolence to anyone when their actions can be more readily explained by incompetence or inattention.

Consistency in sharing information—Do you willingly share information, both good and bad, that is important to your team? Are you honest with them if you do not have the answers? Do you let them know if you need to withhold information and when it can be released? If not, then the rumor mill becomes the primary source of information, often misguided, and mistrust builds among your team. Consistency has many facets.

Walking the talk—As I just mentioned, walking the talk is not a slogan but a first principle in demonstrating integrity. The reputation of a leader is highly dependent on telling the truth, not just what people want to hear. We tend to trust leaders who do what they say they are going to do and, if unsuccessful, take responsibility. If you don't follow through on your word, you most assuredly will develop a reputation of someone who cannot be trusted. While we live in a more complex world today, I can't help but think of my father, who never signed a bill of sale or contract. His mantra was "if my word and handshake are not good enough for you, we shouldn't do business together."

Consistency in location and time—Sharing information is not a one-time event. We are often sharing information to different groups at different locations and situations at different times. The classic example is to listen to the stump speech of most national politicians as they traverse the country, with messages tailored to procure votes. It's no wonder that politicians have one of the lowest trust ratings of any profession. Staff become very suspicious if you change your message based on your audience. Also, changing your message over time and shifting positions will undermine mistrust. Staff look for a consistent sense of direction from their leaders. When changing one's message, great care needs to be taken in explaining the reasons for the change and the anticipated consequences.

Lest this all seems to have little to do with business success, in his book, *The Speed of Trust*, Steven Covey points out the important competitive business advantages in trusting organizations [15].

Empathy—The Willingness to Walk a Mile in Another's Shoes

While there are many attributes of good leaders, the one attribute absolutely necessary to be a great leader is empathy. Most everyone I've met or read about in business espouse the concept of empathy. How often have you heard expressions like "the customer is always right" or "our people are our most important product." The problem is that most of the time these are empty slogans that are preached but rarely practiced. Why is this so?

Once again, we are dealing with the issue of mindset. Most ambitious leaders and managers just don't see a large correlation of empathy and developing relationships with business success. It stems once again from a lack of awareness. There are multiple relationships one can have with another person. In business, the elemental relationship is utilitarian (a day's pay for a day's work; a promotion and raise for a job well done). This is the prevalent mindset of many leaders despite their rhetoric. This attitude can buy a person's arms and legs but not his heart and mind. At a higher level of awareness, very good leaders build relationships based on enlightened self-interest (if I train and develop a person, they can make a larger contribution to my cause or initiative; if I help someone, they may be able to help me in the future). This attitude can buy a person's loyalty.

Finally, at the highest level, great leaders build relationships based not on self-interest but on altruism (a genuine, unselfish caring for someone or some cause with no expectation of return). This attitude builds trust bordering on devotion. Aristotle in his *Nicomachean Ethics* [6] described this type of friendship as virtuous, involving both good character and intellect. The concept is easier to understand when it is related to great secular leaders such as Gandhi, the Buddha, or Martin Luther King. On a scale of 1–10, where 1 is a utilitarian relationship and 10 is an altruistic relationship, where would you rate with your team members?

A great current example of how effective an altruistic leader can be is Arthur T. Demoulas, the CEO of Market Basket, the private supermarket chain. Despite decades of success in building the Market Basket, the Board of Directors fired him. While there were many undisclosed family disagreements, the official reason for his dismissal was essentially distributing too much of the company profits to the employees in the form of generous benefits, including a profit-sharing plan. His firing resulted in an unprecedented workers revolt. Hundreds of nonunion warehouse workers and drivers jeopardized their livelihoods and refused to deliver produce to each of Demoulas' 71 stores, thus crippling the supermarket chain. Employees also led a successful movement to boycott the stores. After several months, the company capitulated and sold the company to Arthur.

What would cause employees to jeopardize their livelihoods and risk being replaced for their fired manager? Here are some of the quotes appearing in the August 27, 2014 edition of the Boston Globe that will shed some light on the matter. "He'll walk into a warehouse and will stop and talk to everyone because he's genuinely concerned about them," said Joe Schmidt, a store operations supervisor. "He cares about families, he asks about your career goals, he will walk up to part timers, and ask them about themselves. To him, that cashier and that bagger are just important as the supervisors and the store management team." Schmidt said. Demoulas once called a store manager after he heard the man's daughter was critically injured in a car crash. Demoulas wanted to know whether the hospital she was in was giving her the best care possible. "Do we need to move her?" he asked. "He is just a good man," Schmidt said. Demoulas is beloved by the workers not only for offering generous benefits—including a profit-sharing plan—but also for stopping to talk to workers, remembering birthdays, and attending funerals of employees' relatives. On my empathy scale described earlier, I would have to rank Arthur a 9 or even a 10.

Like Arthur T. Demoulas, truly great leaders listen, think, converse, and act empathetically a majority of the time. When listening to others, they focus both intellectually on what the person is saying and also emotionally about what the person is feeling (i.e., listening from the heart as well as the head). When formulating plans, they factor not only the potential outcomes but also the feelings and aspirations of others. When conversing with others, they hold their positions gently and try to understand the other's point of view. Finally, their actions demonstrate genuine care for others. You know you are moving in the right direction when your heart warms knowing that you contributed to someone else's success.

Teaching Skills—The Ability to Foster Self-Learning in Others

The last attribute of coaching is the ability to teach. Great coaches are great teachers, not in the grade school sense of imparting knowledge but in the professional sense of investing the time to truly understand the talent, motivation, and aspirations of highly skilled individuals and providing peer-to-peer feedback. The coach's role is to increase their subject's level of awareness and guide them in the process of self-discovery.

4.11

The coach's role is to increase their subject's level of awareness and guide them in the process of self-discovery.

In my experience, coaches that have these teaching skills and motivation are an exception rather than the rule.

Many leaders make the mistake of assuming that they have nothing to teach highly intelligent and educated staff. Nothing could be further from the truth. In fact, in my experience, it is just the opposite. Experts become that way by concentrating and focusing on their area of specialty at the expense of many other skills and behaviors that are needed in today's complex organizational environment. Coaches identify leadership gaps in awareness and behavior to help highly professional staff gain the breath of skills to complement their tremendous depth in their primary field of endeavor. These could include the nine attributes of leadership I've included in the Performance Trilogy, including imagination, courage, persuasion, conscientiousness, productivity, discernment, integrity, empathy, and teaching skills.

First and foremost, good coaches do their homework. They spend time getting to know the strengths, weakness, aspirations, and motivations of their staff. They understand that an important part of their leadership responsibilities is placing their staff in positions to succeed, challenging them with meaningful assignments, and supporting their development and growth. In a new relationship, the better one understands the aspirations of their staff, the more likely they are to guide their assignments and development.

Most of all, organizations recognize the importance of supervision and coaching and have sophisticated annual performance planning and development programs in place. "Management by objectives" has been proven to be one of the more powerful tools in management theory over the past 30 years when successfully practiced. Unfortunately, poor implementation of this practice has many questioning its effectiveness. Most managers oftentimes dread the annual performance review. However, when properly executed, the performance planning and review process are highly effective ways to balance the need for performance (what can you do for the organization?) with the need for personal accomplishment and growth (what can the organization do for you?). Good coaches facilitate this process with integrity, empathy, and dialog. In an upcoming chapter, I will further elaborate on coaching to distinguish between coaching for performance and coaching for development: why these two functions should be conducted separately, and my suggestions for transforming the role of human resources in organizations.

☐ Chapter Summary

There are nine critical leadership attributes that will ensure success.

Leadership Attributes

1. Imagination—The ability to look at what everyone else looks at, and to see what no one else sees

2. Courage—The ability to act despite considerable risks

3. Persuasiveness—The ability to influence

Management Attributes

4. Conscientiousness—The motivation and discipline to be thorough and dependable

5. Productivity—The ability to manage one's energy and focus

6. Discernment—The ability to consistently make good decisions based on cognitive and intuitive judgment

Coaching Attributes

7. Integrity—Behavior that is genuine in words and actions

8. Empathy—The willingness to walk a mile in another's shoes

9. Teaching Skills—The ability to foster self-learning in others

☐ **References**

1. Zeigler, Bernard, Praehofer, Herbert, Kim, Tag Gon, *Theory of Modeling and Simulation,* Academic Press, San Diego, CA, 2000.
2. Hickman, Craig, Silva, Michael, *Creating Excellence: Managing Corporate Culture, Strategy, and Change in the New Age,* Dutton Adult, London, 1985.
3. Machiavelli, Niccolo, *The Prince,* Dover Publications, Mineola, NY, 1992.
4. Robbins, Tony, *Unlimited Power,* Simon and Schuster, New York, 1986.
5. Robbins, Tony, *Awaken the Giant Within,* Simon and Schuster, New York, 1991.
6. Irwin, Terrence, *Nicomachean Ethics by Aristotle,* Hackett Publishing, Indianapolis, IN, 1999.
7. Covey, Stephen R., *The Seven Habits of Highly Effective People,* Fireside (Simon and Shuster), New York, 1989.
8. Hickman, Craig, Silva, Michael, *Creating Excellence: Managing Corporate Culture, Strategy, and Change in the New Age,* NAL Books, New York, 1984.
9. Drucker, Peter, *The Practice of Management,* HarperCollins, New York, 1954.
10. Allen, David, *Getting Things Done,* Penquin, London, 2001.
11. Gladwell, Malcolm, *Blink, the Power of Thinking without Thinking,* Little, Brown, New York, 2005.
12. Welch, Jack, *Jack, Straight from the Gut,* Warner Business Books, New York, 2001.
13. George, Bill, *True North, Discover Your Authentic Leadership,* Jossey-Bass, Hoboken, NJ, 2007.
14. Shaw, Robert B., *Trust in the Balance, Building Successful Organizations on Results, Integrity, and Concern,* Jossey-Bass, Hoboken, NJ, 1995.
15. Covey, Steven M. R. *The Speed of Trust,* Simon and Schuster, New York, 2006.

Managerial Leadership

Leading Teams

5

CHAPTER

Transitioning from Me to We

In Chapter 3, we described a process of deciding whether to take on a leadership assignment based on the magnitude of the challenge and the leadership attributes you bring to it. A key element discussed was examining your motives. The transition from individual performer to team leader may be the most difficult career decision you will ever make. It was for me. Make no mistake, the transition is a transformational one requiring you to evolve an entirely new professional identity. Most management textbooks and courses describe the competencies needed to be a good manager but miss the most important first step: additional self-exploration to develop a new and different mindset that will lead to new behaviors. A good source to help with such personal learning and understanding the roles and responsibilities of management is *Becoming a Manager,* by Linda Hill [1].

> **5.1**
>
> The most important first step in becoming a manager is additional self-exploration to develop a new and different mindset that will lead to new behaviors.

Let's explore the thought processes of the transition, describe the roles and responsibilities of a science and technology (S&T) manager, and conclude with an overall vision and plan of action for the transition.

☐ I Am Excited about the Opportunity of Becoming a Manager

The prospect of being promoted to a managerial position is an exciting one. It is a recognition of outstanding individual performance and an expectation that you could potentially have a larger impact on the organization. This is a pretty good boost to one's self-esteem. In considering such an assignment, there are some positive factors you will initially consider.

- *I can earn more money*—The transition from individual performer to manager is usually a significant promotion with a substantial salary increase. I can better provide for my family and improve my lifestyle.

- *I will gain power and influence*—In my new position, I will be the decision maker and set the standards for my group. My staff will have to listen to me and do as I say. I will be treated with more respect by others both within and outside the organization.

- *I will have more control*—I will be in charge of my groups' mission and get to choose what we work on.

- *I will make important decisions*—I will be able to provide input into the organization's vision and strategy and lead my group in support of the strategy.

- *I have a definitive career path*—There will be opportunities for additional promotions up the management ladder.

☐ I Am Worried about the Challenge of Becoming a Manager

Such a dramatic change in one's career path also brings its share of cognitive dissonance. Your professional reputation and success has been built on producing good science or engineering. You are probably confident about your future as a senior scientist or engineer. You do not know whether you have what it takes to be a good manager. Could you fail? Some negative factors you will consider include

- *Will I lose all of my technical expertise?*
   Being a good scientist or engineer requires 110% effort in keeping up with the literature, doing technical work, and corresponding with peers within and outside the organization. In becoming a manager, I will not be able to commit the time necessary to maintain technical competence.

- *What about the respect from my technical peers?*
   My self-esteem and that of my peers is judged by our technical accomplishments. If I lose technical expertise, how will I be viewed by my technical peers? What could I possibly do as a manager to maintain that level of professional respect?

- *Will I just become a paper pusher?*
   A manager's role as viewed by my peers is mostly administrative. The "real work" is done by technical staff.

- *How can I measure my success?*
   Will my success now depend on the performance of other people? What happens if they don't perform? How can I measure my individual performance if I don't do any "real work"?

- *Will I have to deal with other people's problems?*

  Now that I am responsible for other staff, how do I handle nonperformers?

- *From my observations, a manager's job is highly demanding and stressful.—Is it worth it?*

☐ Managerial Roles, Responsibilities, and Expectations

Many of the positive and negative observations said earlier can be either true or false depending on the circumstances. Before answering some of the questions raised, let's explore the roles and responsibilities of new managers and the expectations placed on them. You will find that when all is said and done, you will be judged by senior management based on three factors.

1. How well did you achieve the performance objectives of your group?

   Your group was formed to achieve certain performance goals and your responsibility is to ensure that it meets those goals. This is primarily the top-down expectation from senior management. Your reputation as a manager will be based on how well you meet those goals. Remember that *managers get things done.*

2. Have you gained the respect and trust of your group?

   Meeting your group's goals is highly dependent on the talent, motivation, and hard work of your staff. As we discussed in Part 1, you will need to ensure that your staff develops faith in your strategy, confidence in your execution, and trust that you have their interests at heart.

3. How well do you get along with your peers?

   As a new manager, you will have the responsibility for providing and receiving input and deliverables to other groups in your organization. You will need to develop mutually beneficial relationships with your peers to accomplish this.

☐ Transitioning to a Management Role Will Be Both Exciting and Challenging

There is definitely much truth to the challenges involved in the transition from a technical to a managerial career. However, many of the impressions that you have probably formed are based on your direct experience with both good and bad managers (mostly bad). Having experienced these challenges personally and through supporting the transitions of many of my technical staff to managerial positions, I feel confident in providing the following answers to the questions raised in the beginning of this chapter.

- *I can earn more money*

  This is generally true. While there are career ladders in most companies for technical staff, the salaries and bonuses earned by successful managers are higher than senior staff scientists.

- *I will gain power and influence*

  This is partially true and depends entirely on you. You have power in that you control the performance evaluation (and salary increases) of your staff. However, your influence is highly dependent on the amount of trust gained through your actions and behavior.

- *I will have more control*

  You will find it most surprising that you will have much less control than you initially might think. Accomplishing goals through others means that you are dependent on them for your performance. The ability to influence performance comes more from your support than control.

- *I will make important decisions*

  This is very true. Your success will be highly dependent on how well you make decisions about the direction of your group's work, the development and satisfaction of your staff, and the relationship you develop with your peers.

- *I have a definitive career path*

  This is mostly true as the management ladder is usually well defined in most organizations.

- *I will lose all of my technical expertise*

  I don't believe this to be true. You will lose some of your technical depth, especially in the hands-on laboratory and engineering procedures, but will gain significant breath. You will be providing technical reviews for all of your staff's work and, from these reviews, gain considerable experience in experimental design, prototyping and scientific and engineering interpretation.

- *I will lose the respect of my technical peers.*

  This is not necessarily true. While over time, you will fall behind your technical peers in the current research trends, you will be gaining valuable experience in creating value from science by balancing your knowledge of science and your organization's business. The ability to help technical staff turn their ideas and hard work into successful ventures will gain you a tremendous amount of respect.

- *I will just become a paper pusher.*

  It is true that there is a lot of administrative requirements of managers. However, by learning new organizational skills you can easily manage to meet these requirements with less than 20% of your time. The best managers *actively manage* by understanding their organization's strategy, translating it in a way that sets the performance objectives of the group, and works to balance their staff's ability to meet those objectives while pursuing their career goals.

- *I can measure my success.*

  This will require a fundamental change in your definition of success. The definition will change from how well did I perform to how well did I build my organization and make my staff successful.

- *I will have to deal with other people's problems*

  This is definitely true. So much so that unless you reframe the question to "I will now have the responsibility to ensure the success of my staff," you will have a great deal of difficulty making the transition.

- *From my observations, a manager's job is highly demanding and stressful. Is it worth it?*

  From my perspective, the answer is definitely yes. There is no more nobler goal than to lead people to excellence, fulfillment, and collective achievement [2]. Having said this, the decision on your career path should be based on your talents and motivation. There is extraordinary satisfaction to be had in pursuing either a technical or managerial career. Since most of you have a pretty good idea of your technical career path in front of you, my intent here is to present a balanced view of a career in S&T management.

> **5.2**
>
> There is no more nobler goal than to lead people to excellence, fulfillment, and collective achievement.

☐ Developing a Managerial Identity and Style

There has been lots written about the roles and responsibilities of managers and the competencies needed to fulfill those roles. I've seen very little written on the dent in self-esteem when making the transition from the height of scientific competence to the bottom rung of the management ladder. It's like starting an entirely new career without any education or training. Even those fortunate enough to return to school and obtain an MBA still worry about how little they know about the art of management.

Learning how to manage has been compared with learning how to ride a bike. You can read and listen to all the instructions in the world about how to do it. Put one foot on the pedal. Push off with the other foot to accelerate the bike. Swing your other foot over the seat and engage the other pedal. Pedal with both feet while maintaining your balance and leaning into the turns. But until you try it a couple of times and fall, you really don't understand what it takes to ride a bike or to manage a group of staff.

I believe that there is both art and science in trying to manage. The science part is well described and can be easily taught. There are excellent management processes for designing a project and putting together a work plan to include a work breakdown structure, staffing plan, schedule, and budget. The art comes in trying to execute a management plan by coordinating your technical staff who all have their own ideas about what should be done and how to do it (more on this subject in Chapter 7). I've often thought that organizations are not complicated. Organizations are simple but run by complicated people.

> **5.3**
>
> Organizations are not complicated. Organizations are simple but run by complicated people.

Another important consideration when taking on your first management assignment is the sphere of influence you have, as described in Chapter 3. It is likely that in your new group leader position, you will be supervising several of your colleagues who consider you their equal or worse still not quite their equal. This is a critical stage of your newly minted position where you will, consciously or unconsciously, develop a new professional and social identity. I've seen dozens of management styles in my travels. The three most common include the need to be liked, feared, or respected.

The most common management style to emerge from newly appointed managers is to remain buddies with their colleagues. After all, the best way to accomplish mutual goals is to get along and not be seen as superior to your colleagues. This includes the Friday night socials and afternoon lunches that are typical of many work groups. Maintaining good social relationships with technical colleagues is not in itself a bad policy and can foster an esprit de corps among the group. The risk that you will take however by being just "one of the guys" is not being taken seriously when faced with decisions that you inevitably will have to make that your colleagues disagree with. It is particularly difficult to conduct a performance evaluation on a "close friend." Like it or not, your relationship with colleagues will change when you assume a supervisory position.

The second management style I have seen unfortunately too often is management by fear. Too many managers let their newly obtained power go to their head. There is a level of insecurity based on fear of failure that requires them to demonstrate their power through intimidation. They need to demonstrate to their staff whose boss. Their decision making is authoritarian with such attitudes as "my way or the highway." This management style, while potentially effective in the short run, is a sure path to eventual failure. Ego is why some things that should happen don't, why other things that shouldn't happen do, and why both take a lot longer than necessary [3]. This is the reason why I have insisted that before you try and lead others, you learn how to lead yourself. By developing better self-awareness, you can unlearn deeply held attitudes and habits that do not serve you well.

> **5.4**
>
> Ego is why some things that should happen don't, why other things that shouldn't happen do, and why both take a lot longer than necessary.

The management style that works best is the one that generates respect and trust from your subordinates (and for that matter, your peers and superiors as well). Gaining respect and trust has little to do with your personality or how close the relationship you have with your staff. The key ingredients of respect and trust are competence, integrity, and concern for others [4]. The Performance Trilogy® framework is a good tool that will help.

To gain the respect of your team, you need to convince them that you have a well thought-out, compelling strategy to accomplish your team's goals and that you are competent enough to lead the execution of that strategy. In your first management assignment, you will most likely be executing the vision of your manager or higher rather than your own. Nevertheless, you still can translate that vision into a concrete strategy for your group and communicate it in a way that generates enthusiasm. You then need to increase your staff's confidence in you by flawlessly executing the strategy and making appropriate adjustments along the way when necessary. To build trust, you must act with integrity in both your words and actions. By setting high standards, starting with yourself and communicating them, your staff should know where you stand at all times concerning outstanding performance and appropriate professional behavior. Your staff should never have to guess what you might be thinking.

Finally, and this is most important for new managers to gain trust, you must act with your staff's interest in mind. There is nothing that you will lose trust more than a manager who takes all the credit when things go well and assigns blame when they don't. Taking the time to have a heart-to-heart conversation with your peers who you now supervise will help you understand their

> **5.5**
>
> People don't care how much you know until they know how much you care.

concerns and uncover ways in which you may be able to help them achieve their personal goals while meeting your group's performance goals. You can't move people to action unless you first move them with emotion. The heart comes before the head. People don't care how much you know until they know how much you care [5].

I would be remiss if I didn't warn you about the other side of being a manager. Instead of resistance from your staff, many will try and be super pleasers. Because of your position, it is important to understand the potential impact of your work conversations with staff and choose your words carefully. A lot of what you say offhand will be treated seriously by some of your staff who aim to garner favor by anticipating your needs and wants. I can remember occasions where my staff spent time and money prematurely initiating actions based on my ideas so as to please the boss. With super pleasers, you also need to be careful when asking for advice on your decisions. Rarely will they speak up if they disagree with what you are planning. You can easily develop a false sense of security when you think that you have a consensus about a decision you are about to make.

The art of management is balancing the requirement of formal authority with the need to establish professional, trusting relationships with your staff. This is entirely situational, and there are no simple answers. Too many managers believe that they are responsible for coming up with all the answers when they should be asking the right questions and seeing answers by having an open dialog with their staff. In his book, *Open Book Management* [6], Jack Stack challenges his managers to honestly answer the following questions.

- What are you personally giving to the people you manage?

- Do you spend as much time thinking about them as you spend thinking about customers, or other departments, or people higher up in the organization?

- Do you share your problems or do you keep them to yourself? Have you asked your people for help in carrying your lead? Do they even know what your load is? Have you told them your critical number?

- Do you yourself operate with an open book? Do you let your people know everything that you know?

- Are you getting the benefit of their intelligence, or do you still think you're responsible for coming up with the answers on your own?

- Do your people know what to do without being told, or do they wait to get a list from you? Is everybody working toward the same goal? Does everybody know what it is? Do you let people figure out the best way to get there?

- Do you know what gets your people angriest? Have you ever asked them about their frustrations and their fears? What keeps them awake at night? Have they told you their critical numbers?

- Have you talked to your people about your own fears and frustrations? Can you let down your guard enough to do that? Are you willing to make yourself vulnerable? Do you have enough self-confidence to take the chance that you might get screwed?

- Most important, if the answers to such questions are no, do you really want to change?

The best managers think of themselves as player coaches [7]. They should be the first on the field in the morning and the last to leave at night. They're available to their players 7 days a week from 8 a.m. to 11 p.m. In the business context, being there on the scene and available is a simple necessity—an if-not-forget-it. Timing is everything. If the manager isn't there when he's needed—to supply the blessing or the go-ahead or the missing piece of the puzzle—his people will lose satisfaction, then interest, and zeal.

☐ Do Your Homework

Although management must be experienced rather than learned from a book or classroom, there are several actions that you can take that can prepare you for a management assignment.

Work for a good one—When asked what one should do to lead a happy successful life, Mark Twain once said, "choose your parents wisely." This quote always pointed out to me the importance of working for and learning from an excellent leader and manager. I had the good fortune of working for some excellent managers but also some very bad ones. By paying attention to how highly successful managers make decisions and treat people, you can model their behavior when you are ready to take on a managerial assignment. Success definitely leaves clues. You can also learn what not to emulate from bad managers.

> **5.6**
>
> When asked what one should do to lead a happy successful life, Mark Twain said, "choose your parents wisely."

This is such an important point that I would go so far as to say that selecting a good manager may be just as important or even more important than selecting a good company to work for. A good manager will spend time to get to know you, your professional talents and aspirations, and provide career guidance and coaching regardless of the career ladder you choose. There have been numerous studies showing that job satisfaction is highly dependent on who you work for. It is said that excellent people don't leave companies, they leave their managers. I know this to be true. Despite the excellent reputation and benefits of one of the companies I worked for, I left that company because I didn't want to work for an authoritarian manager who showed no respect for her people. Seven years later, I returned to that company after she had left.

Prototype an assignment—I had the good fortune of working for a company that had an apprentice program. It was called an associate manager. I was responsible for overseeing the work assignments of a technical group, but the formal authority rested with my section manager. As a player coach, I was able to gain management experience under the close supervision of my manager, suggesting decisions about staff and their project work. I was able to get informal feedback on my management style and effectiveness, without it affecting my performance review. It was a way to determine whether I had the desire as well as the potential to take on a formal management assignment.

Later in my career, I duplicated such a program and invited my senior staff to try out a management assignment for a year. The program was well publicized as a rotational assignment with no stigma attached to either staying in such a position or returning to the technical ladder.

Even if your company does not have such an apprenticeship program, you can prototype such an assignment yourself. Look for opportunities to head up a task force or take on a company pro bono assignment. You will get an opportunity to lead an initiative and test whether you have the talent and/or motivation to become a manager.

Find a good mentor—A mentor can not only accelerate your career but also prevent it from derailing. Consider yourself lucky if you find yourself working for a manager who can also be your mentor; it is quite rare however. Most managers focus primarily on performance and rarely are willing to take the time needed to mentor you (one of the reasons I wrote this book!). Also, it is the rare manager that you can feel comfortable expressing your shortfalls and concerns when he is depending on you to perform. I was incredibly fortunate to have two managers who were also mentors early in my career. I trusted them to constructively point out my many faults and provide career guidance.

There are many mentor sources. You can find mentors both within and outside your company. For younger scientists, I recommend reading the National Institute of Health's (NIH) guide on training and mentoring [8] Two criteria I would look for are success and commitment. You want to be sure that the mentors you choose (and I recommend more than one) are highly successful in the respective fields to be sure that the advice you get is based on their life's experiences. They should have a reputation as developers of people. Also, mentoring can be very time consuming if done correctly and successful mentor's time is quite valuable. Make sure that whoever you select is willing to commit the time needed to support you.

☐ Chapter Summary

Managerial roles, responsibilities, and expectations

How well did you achieve the performance objectives of your group?

Have you gained the respect and trust of your group?

How well do you get along with your peers?

Transitioning to a management role will be both exciting and challenging

- *Will I earn more money?*

- *Will I gain power and influence?*

- *Will I have more control?*

- *Will I make important decisions?*

- *Will I have a definitive career path?*

- *Will I lose all of my technical expertise?*

- *Will I lose the respect of my technical peers?*

- *Will I just become a paper pusher?*

- *Can I measure my success?*

- *Will I have to deal with other people's problems?*

- *From my observations, a manager's job is highly demanding and stressful. Is it worth it?*

Developing a managerial identity and style

Managing by love

Managing by fear

Managing by respect

Do your homework

Choose your boss wisely

Prototype your assignment

Find a good mentor

☐ References

1. Hill, Linda A., *Becoming a Manager*, Harvard Business Press, Boston, MA, 2003.
2. McKee, Annie, Boyatzis, Richard, Johnston, Frances, *Becoming a Resonant Leader*, Harvard Business Press, Boston, MA, 2008.
3. McCormack, Mark H. *What They Don't Teach You at Harvard Business School: Notes from a Street-Smart Executive*, Bantam Books, New York, 1984.
4. Shaw, Robert Bruce, *Trust in the Balance, Building Successful Organizations on Results, Integrity, and Concern*, Jossey-Bass, San Francisco, CA, 1995.
5. Maxwell, John C., *The 21 Irrefutable Laws of Leadership*, Thomas Nelson Publishers, Nashville, TN, 1998.
6. Stack, Jack, *The Great Game of Business*, Currency Books, New York, 1992.
7. Townsend, Robert. *Further Up the Organization: How to Stop Management from Stifling People and Strangling Productivity*. Alfred A. Knopf, New York, 1984.
8. Guide to training and mentoring in the Intramural Research Program at NIH, https://oir.nih.gov/sites/default/files/uploads/sourcebook/documents/mentoring/guide-training_and_mentoring-10-08.pdf.

6

CHAPTER

The Art of Supervision

After much soul searching, you have accepted your first managerial assignment with a title like group leader, section manager, department manager, or the like. While you may have had several technicians working for you as a researcher or project engineer, you now are responsible for a technical group including BS, MS, and PhD staff. You most likely were a staff member of that group and have a pretty good idea about its mission and purpose. Your boss has given you his general expectations from which your annual review, salary increase, and potential bonus will be based. What do you do now?

☐ Take Control of Your Agenda

In my experience, the first several months of your first managerial assignment are frenetic and you can find yourself in a reactive mode most of the time unless you create and take control of your own agenda. Your staff will bring you most of their problems and complaints, probe your competence and motives, and challenge your decisions. Most of their issues will be critically important and urgent in their minds and will expect your immediate attention. It is for this reason that I have emphasized and will continue to emphasize throughout this book the need to focus on strategy first.

It is important to have a clear view of what your strategy is (or for new group leaders, their boss's strategy) and have translated that strategy into performance goals that, if achieved, will guarantee the success of that strategy. If your boss has read this book, he will have spent considerable time with you to make sure that your performance goals were SMART (Specific, Measurable, Achievable, Relevant, and Time bound) enough to support his strategy and that they were highly aligned with your career aspirations. If not, then you need to take the lead and present the performance goals that you believe will achieve your collective

> **6.1**
>
> Take the lead and present the performance goals that you believe will achieve your collective strategy as well as meet your development needs.

strategy as well as meet your development needs. This can oftentimes be a less than smooth interaction, and the performance goals may not meet your mutual expectations without considerable negotiation. I can assure you that if you don't resolve your differences at the planning phase, you will just have put off the inevitable difficult conversation to the end of the year when your performance is being evaluated. It is much better to deal with differences in expectations upfront. This will make the performance review process an easy and less stressful exercise.

Once you have a strong agreement with your boss on your performance goals, you can provide the leadership that your staff needs by articulating the agreed upon strategy and its rational; presenting the group goals that will ensure the success of the strategy; and develop individual staff goals that support the group goals. It is critically important to lead this process with your group to provide direction toward meeting your group goals rather than reacting to their immediate needs.

☐ Get Organized

In addition to having a clear strategy, you will need to dramatically improve how you manage your energy and effort. You will quickly find that the number of relationships, transactions, and requests for your attention will increase exponentially, and unless you develop better systems to become more organized, you will be overwhelmed. It is easy to tell when a new manager has not yet mastered the ability to prioritize his time and process information. Telltale signs include being consistently late for meetings, missing assignment deadlines, not returning phone calls, asking to resend information, and the like.

You may have noticed that I didn't use the term time management. As we discussed in Chapter 4, time management is a misnomer as time cannot be managed. You have zero control of time as the minutes and hours in a day will pass regardless of what you do. What you do have control of is what you chose to focus and prioritize on (being effective) and how well you process information to make decisions (being efficient). It is virtually impossible to stay on top of everything that is going on within your sphere of influence. Managing your energy and what you focus on is the key to reducing stress and becoming highly productive on the issues that matter most. If you want a totally new perspective on this, I suggest you read *The Power of Full Engagement* [1], where you will be convinced that energy, not time, is the fundamental currency of high performance.

> **6.2**
>
> Energy, not time, is the fundamental currency of high performance.

Everything we do and think about has an energy consequence. While the number of hours in a day is fixed, the quantity and quality of energy available to us is not. The ultimate measure of our progress is not how much time we spend but rather how much energy we invest in the time that we have. In general, to be fully engaged, we must be physically energized, emotionally connected, mentally focused, and spiritually aligned with a purpose beyond our immediate self-interest. I find that my peak energy levels occur early in the morning when my mind is racing. I have learned to use my morning time for those activities that require the most concentrated effort and creativity, including reading and reviewing documents and producing reports and other content. I try and schedule most meetings mid to late afternoon when my energy level is lower. I'm a much better listener then. I also take numerous breaks every 90 minutes or so, as renewing one's energy is just as important as using it.

As I have mentioned, as a new manager, you will be inundated with requests for your time from your boss, clients, staff, and colleagues. The first system that you need to quickly develop is a way to prioritize the requests and events in your day. As discussed in Chapter 4, a good tool that helps with prioritizing a manager's time is the Eisenhower Urgent/Important principle. He was quoted in a 1954 speech as saying "I have two kinds of problems: the urgent and the important. The urgent are not important, and the important are never urgent." This "Eisenhower Principle" is said to be how he organized his workload and priorities. To minimize stress, it is important to distinguish between urgent and important activities.

- *Important activities* have an outcome that leads to achieving our professional and personal goals.

- *Urgent activities* demand immediate attention and are usually associated with achieving someone else's goals.

Instead of fighting fires, our time and attention should be spent on executing our strategy. You will see this principle demonstrated in Chapter 10 on managing execution.

Important/Urgent Activities

A top priority should always be your important and urgent activities. Since they are important as well as urgent, they should always be done first. However, upon examining these priorities, how many of them were due to procrastinating those important/nonurgent activities until they became urgent? This is where a well thought-out execution plan, including scheduling, becomes very valuable. I have always found that 1 hour of planning saves at least 3 hours of unproductive time down the road. A second cause of important/urgent activities is unforeseen crises that inevitably occur in the course of business and life. This is why it is important to always save a portion of your time for these unpredictable events. If you continuously have a lot of important/urgent activities, identify which of these you could have foreseen and better plan for them.

Important/Nonurgent Activities

You most likely will find yourself spending the least amount of your waking hours in this quadrant. Ironically, this is where you can most improve your effectiveness. Planning and scheduling the most important activities ahead of time will avoid these activities moving into the important/urgent category, causing unnecessary stress. As mentioned throughout this book, focusing on strategy, staff development, and personal growth and renewal is the best path to meeting your goals. The common myth that you won't have time for fun and personal enjoyment couldn't be further from the truth. *Only the organized can loaf.* By focusing on the important/nonurgent quadrant, you can anticipate and schedule important priorities to include personal development and recreation. An important rule is to never allow important activities to become urgent.

6.3

Never allow important activities to become urgent.

Nonimportant/Urgent Activities

By spending time on these activities, you will burn up valuable time that causes you to miss meeting your important goals. What percentage of phone calls do you answer turn out to be important? I'm guessing a small percentage for most managers. Yet your nervous system screams "answer the phone" whenever it rings interrupting whatever productive work you were doing! A common practice that has proven effective is to schedule a time to return all phone calls and text messages when you are looking for a break. I never return phone calls in which no message is left. Ninety nine percent of the time, these are sales calls. When I was fortunate enough to have my own secretary, I had her answer *all* my calls with the response "Dr. Graffeo is not available right now, I'm his secretary, perhaps I can help?"). Once trained, I found that a competent secretary can handle 90% of all calls by giving out information or referring the caller to a more appropriate person to respond.

Another real-time killer as most everyone knows is the amount of emails or text messages received. For most managers, it is several hundred a day. When anyone tells me that they had an incredibly busy day, I ask them "what did you accomplish?" Answering emails or text messages is not doing real productive work for the most part! While communication is a really important part of a manager's job, there are much more effective and efficient ways to accomplish it. One trick that I found very useful was to provide my staff with some simple rules for tagging their emails using the email priority box.

Top priority!!!—The message is urgent and important. Please respond ASAP

Medium priority!!—I would like a response in 24 hours

Low priority!—Read at your leisure, no response is necessary

This is a good way to use the Eisenhower Important/Urgent principle.

Nonimportant/Nonurgent Activities

Unfortunately, there are numerous mindless activity traps that we all fall into that pretty much waste valuable time. When working, the most precious resource that any organization has is its managers' time. It is fine to spend some time here especially if you enjoy the activity (if the activity gives you pleasure and relieves stress, it may actually be a quadrant 2 activity). The important thing is to consciously manage where you spend your time. I have a habit of alternating my reading from historical fiction to biographies to popular novels and feel that it is a good balance for me. If you are spending more than 10% of your waking hours in this quadrant, you will have a great deal of difficulty accomplishing your managerial work without a great deal of stress.

Once you have mastered how to prioritize what to spend your time on (i.e., being effective), you can then work on how well you process information to make decisions (being efficient). During the course of a day, a manager will receive hundreds of pieces of information in the form of emails, text messages, memos, letters, reports, proposals, news articles, requests from staff, colleagues and supervisors, meeting requests and minutes, and the like.

The ability to efficiently process information is a key organizational skill required for managers (see Critical Managerial Attributes in Chapter 4). Relying on your memory or a simple to-do list is a sure way to get overwhelmed and stressed in your new job. The ability to review, prioritize, file, retrieve, and respond to volumes of information from which to make decisions requires an information management system. While there are many such systems available, I have found David Allen's approach [2], introduced in Chapter 4, to be the most useful. He is a strong proponent of using your mind and memory for having ideas not holding them. To use an IT metaphor, treat your mind as a random-access memory (RAM) and store all the rest of the information in your hard drive. His five-step process is relatively foolproof if you develop the discipline to use it effectively.

1. Capture—All information of interest (professional and personal) should be captured whether it be on paper, computer, voice recorder, or link. Toss or ignore everything else.

2. Clarify—Process and prioritize the information using the Eisenhower principle. Decide if you need to take action. If not, file it with an appropriate tag. If a response is required that takes just a few minutes, do it immediately. If not, place it on an appropriate to-do list. If the response requires multiple steps, set it up as a project with a time commitment and deadline.

3. Organize—Place all the information where it belongs using tabs in a simple program, such as Evernote, Notes, or Reminders. There are many such programs, so choose the one that works best for you. For each list item or project, set a reminder to ping you when it is due.

4. Reflect—Review your to-do list and projects frequently starting with important/urgent, then important items. Keep track of the cumulative level of effort required for a given timeline so as not to take on more projects than you can effectively handle. Anticipate upcoming deadlines so that important items get finished before they become urgent.

5. Engage—Use your system to take appropriate action on commitments based on priority; monitor the cumulative commitment of your time; and review and update on a regular basis. Before taking on any new assignments, check your commitment over that timeline and make sure that you haven't overcommitted. If so, reprioritize and drop the least important activity.

While at first glance, this may seem like more work than it's worth, I can assure you that, once you apply this level of rigor to your time commitments, you will wonder how you ever survived without such an information system.

Bear in mind that leaders are stewards of organizational energy. The Gallup Organization [3] has reported that only 31% of American workers are fully engaged at work and the global number is 15%. This statistic is staggering! 85% of employees worldwide are not engaged or are actively disengaged in their job. Given this statistic, it is easy to see that managing employees' time is a waste of effort. Increasing the energy of employees by the methods discussed in this book will dramatically improve both you and your organizations' performance.

> **6.4**
>
> Eighty five percent of employees worldwide are not engaged or are actively disengaged in their job.

☐ Focus Most of Your Attention on Your Staff

In Chapters 10 and 11, we will be discussing the importance of implementing a disciplined performance management system and leadership coaching at an organizational level. Some of the same principles should be practiced at every level of the organization. Your first tendency will be to focus majority of your attention on the tasks at hand to achieve your performance goals. The vast majority of new managers and unfortunately many senior managers make this mistake. Of course, task management is important. It's what your boss expects of you and how your performance will be evaluated. But remember that you are transitioning from me to we and the majority of the work that underpins the successful completion of your goals will be done by your staff. So, the best advice I can give you is to focus most of your attention on your staff.

> **6.5**
>
> Managers are stewards of organizational energy.

Over the years, I have come to believe that there are five key success factors for all managers (Figure 6.1).

1. Surround yourself with talented and motivated staff aligned with your vision and values

2. Set high performance standards starting with yourself

3. Critically, honestly, and frequently evaluate performance

4. Coach up your high performers

5. Marginalize your low performers

Surround Yourself with Talented and Motivated Staff Aligned with Your Vision and Values

The foundation of a manager's success depends on the quality and productivity of the staff that reports to him. If you want to grow your organization, grow your staff. No matter

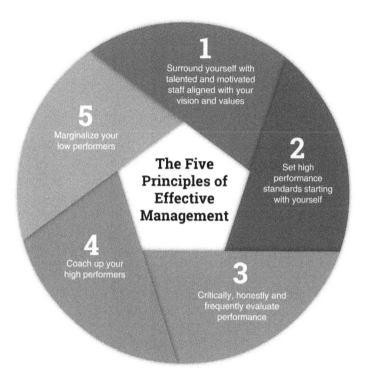

FIGURE 6.1 The five principles of effective management.

how good or how successful you have been as an individual performer, your success now depends on the people you surround yourself with. It is likely that you will inherit staff in taking on your first management assignment. Over time, you will have the opportunity to hire new staff.

6.6

If you want to grow your organization, grow your staff.

It is imperative that one of your first initiatives is to get to know your staff in depth. Too many managers have the mindset that staff are resources to be used as needed to achieve their goals (hence the inappropriate term human resources). The correct mindset is that staff are assets not liabilities, a form of human capital, and like financial capital, should be invested in so as to receive a return on investment. When you view staff in this way, you are always looking for ways to best develop them and put them in positions to succeed. You should be telling your staff that you want to help them get promoted. When they do, they will be able to contribute at a higher level in the organization.

6.7

Staff are assets, not liabilities; a form of human capital, and like financial capital, should be invested in so as to receive a return on investment.

First and foremost, select people who have talent. Talented people have the potential to develop and nothing returns a better investment than talented staff. It is common to focus on experience when hiring or selecting staff for an assignment. This can be deceiving. A person can claim to have 10 years of experience only to find out that he really has an year's experience repeated ten times. While experience is important (after all, the job needs to be done), I would choose talent over experience every time. A person with talent but less experience will start with a higher learning curve, but he will overtake the more experienced person every time in completing the assignment. A good way to view yourself is as a talent manager. Think like a talent agent for performers (singers, actors, etc.) and match the specific talents of your staff to the assignments that need to be accomplished.

While talent is essential for success, it is not sufficient. It is important to access what motivates the individual and whether that motivation aligned with the vision, strategy, and values that you know are important for success. Highly skilled scientists and engineers are usually self-driven, and you need to determine whether their aspirations are aligned with the direction that you want to take your group. I would look for at least 70% alignment between an individual's personal vision and strategy and yours for the group. If it is much less than that, you should seriously question whether the person will be successful working for you. I am reminded of what Lou Holtz, the coach of the Notre Dame football team, said facetiously when asked how he motivates his football players. He responded, "I find the ones that aren't motivated and cut them!." Determine which of your staff align well with your vision and strategy so that you can spend your time investing in them rather than swimming against the current.

Finally, it is important to not only determine early on whether a staff member can successfully complete his goals but also important how he does it. Once again, managers undervalue the importance of culture and behavior in building a successful and cohesive group. An important element of staff satisfaction is the working relationship the staff have with their peers. If your group is populated with some staff that do not share the values that you have espoused for the group, overall performance will suffer. I once heard a manager describe a review that he had with an excellent performer who was a tough task master. "The result of your work was excellent, but I feel like I have to send in a MASH unit to stop the bleeding and dress the wounded!" Jack Welsh had a philosophy that, if a manager who espoused the company's values didn't meet his goals, he oftentimes was given a second chance, but there was no place for a manager who violated the company's values despite meeting goals.

Set High Performance Standards Starting with Yourself

A person's success is directly correlated to his standards. Make sure you set high standards, both for performance and behavior, as a first priority. It is very difficult to raise standards once expectations have been set. When taking over the management of a new group, the first order of business is to set the standards that you expect you and your staff to meet.

> **6.8**
> A person's success is directly correlated to his standards.

Before you begin to set standards for your staff, start with your own standards. The quality of work, speed of progress, and culture of the organization starts with and is highly dependent on the standards of the organization's leader. A best practice is to share your performance goals that you have negotiated with your manager with your direct reports. Point out how these goals are congruent with your personal development goals and are SMART enough to easily evaluate during a performance review. Also point out what you are delegating to your staff and how their performance will directly contribute to you meeting your goals. Being this transparent will garner additional trust among your staff and a willingness to collectively work toward achieving your team's goals.

Just as important as sharing your performance goals, you need to live by the values that you have established. As a role model, do not underestimate the enormous effect you have on the culture or your organization. If your organizational values include respect for others, you cannot show disrespect ever. If it is work ethic, you should be the first person to arrive at work in the morning and the last to leave. If it is having fun, you should support any effort to enjoy work. You get the picture. Your staff will adapt their behavior to the culture you want only if they see you living the values. John Wooden the famous basketball coach is quoted as saying "The most powerful leadership tool you have is your own personal example."

You will see that a recurring theme of this book is that managers should spend a great deal of time building effective relationships with subordinates to understand their needs and aspirations before establishing performance goals. How much time should a manager spend with each person? As long as it takes to evaluate his talents, motivations, and aspirations. This is the single most difficult task to accomplish and also the most likely success factor of a good manager. Why? Employee engagement and energy levels will go up and productivity will dramatically increase. If you practice this one fundamental, I predict you will be in the top five percent of all managers in your organization.

The first step in setting performance standards is your staff's personal development goals. By challenging the staff to be the best they can be within the constraints of their talent level, you can determine how best to develop that person's personal goals consistent with your organizations goals. This is performance coaching at its best. The most fun is to coach staff that have more talent than they are aware of. Increasing their awareness and helping them to start a journey of lifelong learning is extremely rewarding and satisfying to a good manager. On the other hand, coaching staff that believe they have more talent than they actually possess is a difficult and unwelcome chore. It is necessary however for the good of both the organization and the individual. Remember that your role is that of a talent manager. If you put an individual in an assignment that you know they are not ready for or are incapable of executing, the fault is on you, not the staff member.

You will find that by setting personal development goals first, setting organizational performance goals becomes much easier and less confrontational. If done correctly, you will have developed a level of trust that you have the individual's interests at heart in addition to the organization's. You will then find more cooperation working on a performance plan. Always be sure to match the degree of difficulty of performance goals with the specific abilities of each staff member. While stretch goals can be motivational, setting goals that go beyond a person's ability will cause a great deal of stress and result in failure. On the other hand, goals that don't challenge talented and ambitious staff can result in boredom and job dissatisfaction. For a more detailed discussion of what motivates employees, I recommend Csikszentmihalyi's book, *Good Business, Leadership, Flow and the Making of Meaning* [4]. By working with the staff member, you can mutually set the most appropriate performance goals that meet both the organization's and individual's goals.

When establishing performance goals, it is important that they be SMART. As previously mentioned, it is an acronym that stands for Specific, Measurable, Achievable, Realistic, and Time based. It never ceases to amaze me how few performance goals that I have reviewed are very SMART. As a result, the performance reviews and the end of the year oftentimes are confrontational because of lack of specificity and metrics.

When you agree on a well-designed SMART goal with your staff, they can self-review their performance with little more than concurrence from you.

Critically, Honestly, and Frequently Evaluate Performance

Evaluating performance is one of the most difficult tasks that a manager has to perform and dreaded by most young managers. For high performers, it looks easy, a lot of fun, and won't take much time. For low performers, it is an onerous task as no one likes to tell someone they are falling down on the job. It also appears that it will require a lot of time to reply to objections and debate your position. Instead of viewing the performance review process as a necessary chore, I am convinced that it is one of the most important activities that a supervisor can do to ensure both high performance and staff satisfaction. I am going to describe a process that will help you achieve the earlier two results, high performance and staff satisfaction.

First and foremost, ask yourself the following questions.

"What is the best way to exercise my authority?"

"Should I try and get results through love, fear, or respect?

"What can I say to improve my staff's ability to meet their goals?"

"How can I accomplish this while maintaining my staff's satisfaction with their job?"

I have found that reflecting on these questions is the best way to transition from me to we. How you answer the questions will determine your management style and effectiveness. If you seek to manage by love, i.e., having your staff view you as a nice guy and a friend, you will not be honest or critical in your evaluations for fear of upsetting your staff. This leads to a false sense of accomplishment by underperforming staff members and resentment from high performers who see the evaluation system as dishonest and unfair. This results in goals not being met as well as staff dissatisfaction; not very good outcomes!

If you seek to manage by fear, i.e., having your staff view you as a tough guy who carries a big stick and doesn't tolerate mistakes, you can achieve limited success. Fear of no salary increase, being passed over for promotion, or getting fired can be a powerful motivator. However, by adapting this style, you will be guaranteeing that the performance of your staff will be less than optimal. This kind of behavior can only control your staff's arms and legs, and not their hearts and minds. Once again, you will not achieve your goals beyond the minimum acceptable level nor the satisfaction of your staff.

The best way to exercise your authority is by critically and honestly evaluating your staff's performance, thus earning their respect. As mentioned in the previous section, if you worked hard on developing SMART organizational goals and they are aligned with your staff's personal development goals, you can now take on the mindset of a performance coach rather than a boss. If you were coaching a professional athlete or a marathon runner, they wouldn't accept being ordered around or being babied. You would only gain respect as a performance coach if you observe their performance *frequently* and provide critical and honest feedback on what they were doing well that enhanced their performance and also what they were doing that detracted from their performance. Why would you not adopt the same approach and attitude with your "science and engineering" athletes that need to perform at a high level to satisfy their own aspirations and meet organizational goals?

6.9

The best way to exercise your authority is by critically and honestly evaluating your staff's performance, thus earning their respect.

One of the most important goals you should have as a supervisor is to encourage initiative. It is also important how you respond to your staff's ideas and handle their mistakes and failures. While senior executives like to talk about the importance of managing innovation today, very few back up their words with appropriate behavior. Nothing stifles a scientist's or engineer's motivation more than having their ideas turned down. You should always leave some room in performance plans for an individual to propose his own organizational goal and include it as a stretch goal. (This is much better than just increasing the level of an existing goal).

I can remember a manager of mine who was very enthusiastic about submitting a proposal to a government agency for a highly competitive contract. The cost of proposal preparation was quite high, and I felt as though our chances were slim to none. Our organization had already submitted a proposal to that agency during that years' proposal cycle, and it was highly unlikely that this agency would award two contracts to the same organization. Nevertheless, I weighed the financial cost of the proposal with the cost of dampening a highly valued scientist's enthusiasm and choose to fund the proposal. Much to my surprise and delight, we were awarded both contracts, a first for that agency. While I was very happy with the outcome of my decision, I still felt that, even if we had lost the proposal bid, it would have been worth the money to maintain that manager's enthusiasm for his work.

In another example, I personally worked for over a year on developing a business plan for a new business for my organization while also managing and overseeing the projects of more than 500 scientists and engineers. This was an initiative that I was personally committed to and put in a lot of effort on nights and weekends to put together the best proposal I could for my company to invest in. After a year's worth of effort, I presented it to the CEO and investment team.

My proposal was rejected, and I left the meeting quite depressed. Immediately thereafter, I was called into the office of one of the senior vice president who was on the committee whom I didn't know very well. He thanked me for all the effort I had put into the proposal. He let me know that the organization needed more managers like me who were willing to go the extra mile to grow the company and not to get discouraged. "The company refused your proposal but didn't reject you. Bring me some more ideas, and we will seriously consider them." Even though my proposal was not funded, I left his office feeling really good with a healthy respect and trust for this manager who knew how to say no correctly. Sometimes it's better to be kind than to be right.

Coach up Your High Performers

Most managers complain about the time they have to spend with low performers. It seems the squeaky wheel always gets the oil. On the other hand, many are proud of the fact that "I hire good people then leave them alone." Although counterintuitive, you should be doing just the opposite. That is spending much more of your time with high performers than low performers. After managing for some time, I have come to the conclusion that the most precious resource in any company is a manager's time. The manager's initiatives will drive the organization, and it will move only at the speed of the person at the top.

6.10

The most precious resource in any company is a manager's time.

If you spend your time trying to coach up low performers you may be able to improve their performance by 20%–30% at the most with a lot of work. However, there is ample evidence that if you spend your time coaching up your high performers, you can double or triple their performance. Given the value of your time, doesn't it make more sense to spend it where you can create the most value? The key term here is coach, not manage. There is a common misconception that your best scientists and engineers want to be left alone to do their work. This is certainly true if they are being constantly harassed about the progress of their work and meeting their goals and deadlines. As I have clearly stated many times, micromanagement is a poor technique for managing top professionals and counterproductive. On the other hand, by spending more time and paying attention to your top performers as a coach, you can identify ways to dramatically improve their performance while increasing their enthusiasm. Everybody, even the best and the brightest, like to be paid attention to. Always remember, when managers stop paying attention, people stop caring.

6.11

When managers stop paying attention, people stop caring.

As you have learned in the beginning of this book on the Performance Trilogy®, it is most important to engender trust from your top performers by demonstrating that you are interested in their success and that you care. The time that you spend with them initially should be to get to know them better; their strengths, weaknesses, motivations, frustrations. The better you know your top performers, the more likely it will be that you can support their professional growth and performance. Once again, your success is highly dependent on the output of your top performers. Doesn't it make sense to focus more of your time leveraging their performance?

There are several ways that you can support your top performers in a way that is supportive rather than intrusive. The easiest way to start is to eliminate organizational barriers that inhibit their performance and cause frustration. I like the football sports metaphor of the blocking fullback, whose job it is to open up holes in the defensive line so that the star running back can gain the maximum amount of yardage. Find out what these barriers are and figure out ways to eliminate them. They could be administrative processes, staff issues, and unproductive assignments, to name a few. Now that you have developed a team leadership mindset (transitioning from me to we) and recognize that you are in the execution phase of the Performance Trilogy, you can take your management hat off and put your coach's hat on and be the best fullback you can be.

One of the best compliments I ever received was a sincere thank you from one of my top performers years after I had left the organization. He told me that he never appreciated how much I ran interference for him from all the unproductive administrative tasks and micromanagement from senior management until I had left the organization. He said that is productivity and motivation plummeted under the new management.

As a coach, focus your attention first on development rather than performance. This is where the biggest gains in performance come from. By focusing on development first, that is what the organization can do for the high performer, you continue to build trust as well as focus attention on self-improvement, a win-win. I have found that most high performers have specific talents and strengths that are keys to their performance. They also have significant weaknesses. Unfortunately, most managers are taught to have all performers focus on improving their weaknesses in their development plans. By focusing time and energy on multiplying their strengths rather than incremental improvement of their weaknesses, performance can improve by orders of magnitude rather than 20%–30%.

Encourage your top performers to spend just enough time to develop minimum competency in their weak areas and spend the majority of their time building on their natural talents to maximize their strengths. Find creative ways to support your top performer's weak areas whenever possible. You can choose more appropriate assignments, add complementary staff to the project team, and provide support staff to relieve the administrative burden. Some of my most talented staff needed such support to maximize their performance. I needed to support a brilliant scientist with extraordinary skills in experimental design and writing skills with a seasoned project manager with budgeting and scheduling skills. One of my most experienced engineers with a knack for invention needed support from proposal and report writing team members. I assigned a scheduler to a superb public speaker lacking control of his time. This allowed my top performers to focus on what they excelled at and enjoy their work, rather than spending emotional energy and wasting time on their weak areas.

Marginalize Your Low Performers

Despite your best efforts to recruit, hire, and train staff, it is inevitable that you will have to deal with low performers. The worst thing you can do is nothing. Ignoring poor performance will not only reduce the overall success of your team but will result in demotivating your high performers. Everyone wants to work in an environment that favors meritocracy. There is no faster way to lose respect from your top performers than being viewed as a manger that tolerates low performance or worse still has favorites that are untouchable.

Before taking any action, it is imperative that you determine the root cause of poor performance and whether it is fixable. Three areas to investigate are lack of motivation, undeveloped skills, or insufficient talent. Oftentimes, poor performance can be attributed to a lack of motivation for a number of reasons, both professional and personal. The first thing to check is whether the job function is either too difficult, leading to anxiety and paralysis, or too easy, leading to boredom. Remember that as a talent manager, one of your key role is to place staff in assignments where they can succeed. Oftentimes a staff member is reluctant to change positions or jobs even though he knows that he is not happy in his current job. It requires you to take him out of his comfort zone. If the job function is a mismatch for the staff member, you need to find a different one within the company if available or suggest a move to a new company that is a better fit.

There are many times that a lack of motivation is due to personal reasons. Asking someone to leave their personal lives at the door when they come to work shows a clear lack of interest in the staff member and shows poor management judgment. As a good coach, finding out the root cause of poor performance and recommending solutions is your job. There are many personal problems (marital, financial, health, substances of abuse, etc.) that inhibit a staff member from fully engaging in his job. Finding out whether the problem is short term or chronic and whether there is an appropriate support system should be your first action item. You can then choose some support options available to you, such as leave of absence, or professional help.

If motivation is not the problem, it could be a lack of skill or proper training. If there is a strong desire on the part of the staff member to improve, it may be worth your while to get him the training needed to solve the problem. If the training does not work, it is likely that the staff member just doesn't have the necessary talent to perform the job well. You then need to find a job that more appropriately matches his talents. Lacking such a position, you need to have the staff member leave the company.

In the spirit of spending most of your time with high performers, you need to resolve problems with poor performers quickly. After some initial coaching sessions to form a course of action, you may want to delegate the execution to your human resources professional.

☐ Chapter Summary

Take control of your agenda

- Develop a clear view of your strategy

- Translate your strategy into SMART objectives that, if accomplished, will lead to success

- Communicate your strategy, objectives, and priorities to your staff

Get organized

- Manage your energy, not your time

- Prioritize your important and urgent tasks for effectiveness

- Use a system to process information to make decisions efficiently

Focus most of your attention on your staff

- Surround yourself with talented and motivated staff aligned with your vision and values—The foundation of a manager's success depends on the quality and productivity of the staff that reports to him

- Set high performance standards starting with yourself—The quality of work, speed of progress, and culture of the organization starts with and is highly dependent on the standards of the organization's leader

- Critically, honestly, and frequently evaluate performance—Take on the mindset of a performance coach rather than a boss.

- Coach up your high performers—If you want to grow your organization, grow your staff

- Marginalize your low performers—Determine whether the cause is lack of motivation, undeveloped skills, or insufficient talent.

☐ References

1. Loehr, Jim, Schartz, Tony, *The Power of Full Engagement*, Simon and Schuster, New York, 2003.
2. Allen, David, *Getting Things Done*, Penquin, New York, 2001.
3. Gallup Organization, State of the Workplace Report 2017.
4. Csikszentmihalyi, Mihaly, *Good Business, Leadership, Flow and the Making of Meaning*, Penguin Putnam, New York, 2003.

Project Leadership

☐ The Transition to Project Leadership

Successful first-line supervisors often advance their careers when asked to take on large company cross-functional projects. This will be the first time that most managers will be leading professionals with different backgrounds and expertise toward a significant company goal. This could be the submission for approval of a new drug to the FDA, writing a complex proposal to a new client, building new science and technology (S&T) facilities, and the like.

The traditional view of the project manager's role, practiced for decades, is to deliver high-quality products and services that satisfy customer requirements on time and within budget [1]. These are the principles that are taught in most project management courses and are what I call the science of project management. This traditional role is important but insufficient to satisfy today's demanding business requirements.

Successful companies today are developing stronger relationships with their key clients; emphasizing value-added services over products; reengineering their critical business processes to improve productivity; and demanding a higher, more consistent return on their assets. Traditional experts in project management no longer have the luxury of focusing the majority of their efforts on delivering the product/service on time and on budget. Today's project managers need to do much more, i.e., become project leaders and not only practice the science but also the art of project management. By distributing the leadership role further down the organization, senior management can better horizontally align the entire organization to meet its goals [2].

There is no question that project management is a complex and demanding profession. The expanded roles that I will describe makes the job even more challenging. Clearly, today's project manager needs a variety of tools and techniques to help make the job more manageable.

Traditional books on project management have emphasized the science of project management. These are excellent treatises on developing a work breakdown structure and assigning responsibility, schedules, levels of quality, and deliverables to the individual work packages; then integrating these work packages using simple Ghant charts or more complex critical path analysis. These are important tools and can make a project manager's job easier.

The emphasis of this chapter is not on the science but the art of project management. While teaching courses in project management, I always introduced the course by asking each participant to list one outstanding project and one problem project that they have been involved in and the reasons why they were labeled such. I then characterized the results in term of whether they are related to the art or science side of a project. With a sample size of over 500 projects, greater than 80% of the successes and/or failures were attributed to the art of project management or project leadership. This also tracks with my personal experience.

The art of project leadership deals with subjects that traditionally technical professionals find uncomfortable or feel is unnecessary but more often than not drive the success of the project. These include organizational culture, teamwork, communication, troubleshooting, and conflict resolution. While at first glance, these may sound situational, there are underlying factors that lie at the heart of most situations. Understanding why these factors drive most projects today and understanding the principles of leadership when dealing with multiple stakeholders is the purpose of this chapter.

The purpose of this chapter is to present a new model of the role of the project leader in today's business environment based on the principles of the Performance Trilogy®. My belief is that the project leader is central to the success of today's service-oriented, S&T-based businesses. Putting these fundamentals to work on your projects will lead to a higher likelihood of project success, career success, and personal satisfaction with less stress.

> **7.1**
>
> The purpose of this chapter is to present a new model of the role of the project leader in today's business environment based on the principles of the Performance Trilogy.

My confidence in these bold claims comes first from the adoption and practice of these principles over the past 40 years as a project manager on government projects (large, medium, and small) as well as industrial projects with the pharmaceutical, chemical, and petroleum industries. Second, as a line manager and consultant, I have had the pleasure of participating in the celebration of successful projects and the pain of conducting damage control on projects gone awry and all the lessons learned that go with such participation. Finally, most of the principles in this book have been tested and challenged by over 1,000 senior managers who have taken my project leadership courses at companies I have worked for and consulted with.

While this chapter is written primarily for scientists and engineers in a consulting or contract research environment, the principles are more widely applicable. For example, there are substantive discussions on serving external clients, which involve fee for service. The principles are equally valid when serving "clients" within your own organization. Also, many managers without technical backgrounds have provided me with positive feedback on the applicability of these principles to any professional service project.

☐ The Importance of Project Leadership

Project leadership is the most important business process being performed in today's organizations. While at first you may question this bold assumption, I hope to convince you of its validity by describing how much of today's business relies on the execution of projects and the central role played by the project leader.

Approximately eighty percent of today's businesses are service businesses [3]. This represents the business-to-business transaction. When you take into account the internal projects and tasks conducted within a company in support of the final product or service delivered to a client (the R&D project, the pilot-scale project, the market survey project, the Beta test project, etc.), it is clear that today's business for the most part depends on project execution for its success.

The latest management experts have been focusing on the new service economy and the changes in management that are necessary to survive and compete in this new business environment. Business process redesign [4,5] has been the target of much criticism because of its association with massive layoffs. In reality, this management practice represents a very useful tool to evaluate and improve the way in which companies organize their internal processes (and the projects that support them). The emphasis is on horizontal integration in support of the final product or service. There is much more emphasis on project teams.

The total quality management movement that swept the business community in the early 90s emphasized customer supplier relationships [6,7]. The concept was that every professional, project team, or department within a company was a supplier with responsibility for the delivery of a product or service to a customer. At the same time, this team was a customer of several suppliers. This chain of customer–supplier relationships made up the value chain, whereby the final product or service was delivered to the external paying customer. The concept of stakeholders was also introduced at this time. It was recognized that a project team might have more than one customer that they had to satisfy in delivering their service. Understanding how to satisfy multiple stakeholders in today's complex business environment became a necessary complication.

Finally, the increase in sophistication of today's buyer of services, brought on by the easier access of information and the multiple choices available, has led the marketing experts to rethink the concept of strategic marketing. The focus today is less on market share and more on identifying specific market niches that you are capable of dominating through superior service and strong client relationships.

The basic premise of this chapter is that today's businesses and organizations in general execute through projects. Even product-based businesses derive a great deal of their productivity through project-based activities. Whether it is a market survey, sales campaign, advertising program, R&D project, consulting assignment, training program, or travel project, project teams under the leadership of a project leader determine success. As in the previous sections on personal leadership and the upcoming sections on executive leadership, the principles of the Performance Trilogy have been incorporated into the project management fundamentals presented in this chapter. While the model is applicable to a wide variety of professional service projects, most of the emphasis in this chapter is placed on technical projects conducted in S&T-based businesses.

Before describing the project leader's role in detail, it is important to define a technology-based business and the customer service model. We will then focus on how a project leader can use the fundamentals in the Performance Trilogy to maximize project and company performance.

☐ Technology-Based Business

First, let me define a technology-based business on which the concepts of this chapter are based. A technology-based business is one that has products and/or services that require significant technical input and advice to ensure client satisfaction. This includes most consulting and contract research companies, companies that include system integration and technical support with their products, and most R&D and product development departments within companies. It does not include businesses that require little technical advice, even though the product may have a high degree of built-in technology (such as the purchase of an automobile or cell phone). The concepts presented in this chapter are also useful for professional service companies that have little technical content (public relations, marketing, legal, management consulting), but the examples and anecdotal stories in the chapter are geared toward the technical professional.

Business Development in a Technology-Based Business

Since business development has different meanings in different organizational settings, let's define business development more generally as both the marketing and client relationship development necessary to acquire new clients. How one markets and manages a technology-based business is distinctly different than an off-the-shelf product/service business, and the role of the project leader is infinitely more important.

Successful technology-based businesses command higher margins due to their ability to differentiate their product/service by providing higher value to their clients. Many companies are developing strategies to migrate toward value-added services to improve their competitive positions and profits. The path to this success however is rocky since the onus is on the S&T-based company to convince prospective clients of this value before delivering it! [8]. To complicate matters even more, initial success of a technology-based product/service with early adopters is difficult to sustain with mainstream clients without a substantial technology transfer step. This process has been described in crossing the chasm [9].

An S&T-based business relies much more heavily on the project leader for its success with clients. A prospective client with no previous experience with your company gets no upfront guarantees on your company's product/service, like he would when buying a car or a cell phone. In signing a "contract" (whether with an external or internal client), the client gets a promise of some future service that he believes will be valuable to him if delivered as described in a proposal. This requires a great deal of trust and confidence in the person that your organization is proposing to deliver on that client promise, i.e., your project leader.

Existing clients more often than not continue to do business with your organization because of the relationship that has been developed with your project leader. Surprisingly, this is the case regardless of how the project went, so long as the results were acceptable, and the project manager has developed the trust of your client! Oftentimes, the chief reasons for a client selecting (i.e., taking a chance on) your organization is due to referrals (i.e., testimonials from current clients) and performance of similar work with another department or a colleague or trusted friend in the industry.

The tried and true expression "the best marketing in the world is a job well done" still holds true. Experienced marketing and sales professionals lead with this strength in an S&T-based company.

Staff Development in a Technology-Based Business

As was the case in business development, a company's approach to the development and management of professional staff also requires special consideration. In reading every company's mission statement, invariably one will find a statement such as "our staff are our most important asset." Unfortunately, actions speak louder than words and many companies just give lip service to their employee commitments. In companies with strong off-the-shelf products, the leverage that employees have on customer sales and satisfaction is not exceptionally high, and companies can get away with employee neglect for reasonably long periods at a time before it eventually catches up with them.

In a technology-based business, however, failure to put employees first is tantamount to business failure. Since technical input and advice is an essential part of the product/service itself, your people literally are your product. A technically competent and motivated staff is your most important competitive differentiator (yes, even more important than the product/service you are selling!). In a technology-based business it is easier to find new clients, technology, and financing than to find and replace key technical staff.

7.2

In a technology-based business, failure to put employees first is tantamount to business failure.

The implications for staff development are obvious. Technically competent staff are motivated by challenging assignments, the opportunity to grow professionally and technically, and recognition for a job well done. They want to report to someone whom they respect professionally and who are interested in their development. A big complaint is that there is never enough time for training and professional development, because staff are too busy on their projects. They understand that keeping up with technology is key to their job security and value a company that has the same beliefs.

How often have you heard from well-intentioned managers or human resource professionals that one of these days when our work lightens up, we are going to do some training. No matter how well intentioned, it seldom happens and for good reason. The economics of S&T-based businesses require high utilization rates to keep their overheads competitive and make a profit. The economics of the business dictates that 80% of a professional's working hours is spent working on the company's projects for both internal and external clients. Administrative and marketing support duties leave precious little time for professional development and training. Nonproject activities, no matter how valuable they may seem, are viewed as activities that reduce profit. That's a powerful obstacle to training!

In discussions with hundreds of technical staff over the years, invariably their experience has shown that the training that they received while working on a few key projects (i.e., on-the-job training) was the most relevant to their advancement and career success. Formal courses and training sessions not reinforced by their day-to-day project work just didn't have a lasting effect. Here's the eureka moment. If the most useful professional development takes place on project and the majority of a professional's time is spent working on client projects, wouldn't it make sense for an S&T-based company to build professional development into their client projects? There does not have to be any increase in the fees to clients, because the company budget for training has already been built into the overhead charged to clients. The problem is not with the budgets but with appropriate implementation.

In other words, by shifting the training budgets and responsibility to the project manager, there is finally a path toward professional development that does not conflict with the company's earnings goals. However, the principles of the Performance Trilogy require the transition of the project leader from manager to coach.

> **7.3**
>
> By shifting the training budgets and responsibility to the project leader, there is finally a path toward professional development that does not conflict with the company's earnings goals.

Business Management in a Technology-Based Business

The next topic to discuss on the importance of project management is the business of business. Traditionally, this lonely role has been exclusively served by the line manager or supervisor. Once a year, he would closet himself in his office and work continuously for days on next year's budget. While he would ask for input on occasion from various project managers and staff professionals, the assumptions, calculations, and conclusions were pretty much his alone.

In January, he would triumphantly present his budget for the year and begin the process of comparing monthly actuals to budget. He also carried the burden of working capital, cash flow, and capital budgets pretty much himself. It is no wonder that most professional staff, having no input to the process, felt no ownership in the budget and furthermore didn't relate how his everyday activity contributed or not to the budget. The reason that is given for closely holding the budgeting and other business processes is that most line managers truly believe that their professional staff are business illiterate. I find it humorous that many such managers are labeling technical staff who have successfully mastered complex subjects such as thermodynamics, quantum mechanics, and differential equations, as incapable of mastering arithmetic!

There is overwhelming evidence to the contrary. I question the assumption that technical staff are not interested in the process of making money and their only satisfaction comes from performing their professional tasks. Today's technical staff who are labeled in work as business illiterate are the same professionals that are oftentimes managing a complex financial portfolio at home and measuring results versus the S&P 500!

First introduced by Jack Stack in his book *The Great Game of Business* [10], business literacy training at all levels of the company can unleash a level of enthusiasm and entrepreneurship previously unheard of in all but employee-owned companies. The implications for the project leader are profound. He becomes the central figure in the second fundamental of the Performance Trilogy execution, by being more involved in financial management, including project cash flow and client receivables; asset management, including capital equipment and facility employment; and risk management that protects the company's reputation. More on this topic is provided later in the chapter.

The Customer Service Model

We live in a world of instant access to technology and markets. This means that competition is rampant. While it has always been true that business development is the lifeblood of any company, today, more than ever, "nothing ever happens in business until somebody sells something." Traditionally, S&T-based businesses have focused on their products. Development and delivery of products drove companies' strategic plans. In these "product-driven" companies, marketing or project development was just one of the many business functions and not the most important one.

Today, the best S&T-based companies have learned that it's not good enough to focus on their products. Instead, they focus on their markets and customers. These "market-driven" or "customer-focused" businesses understand in depth that it is the customers, not the products, that allow their companies to exist and flourish. Today's best companies have found that "value-added" services are necessary if they want to stay in business. What are these value-added services, and how do they change how a company does business? The answer is in the customer-service model.

I have embraced the customer-service model for conducting an S&T-based business in a rapidly changing world. In this model, there are three critical business functions that a technology-based company must embrace to be successful: customer-relationship management, service delivery, and continuing technology development. Also, there are three roles that project managers working in the market-driven or customer-oriented business must learn: building client relationships, developing talented and motivated staff, and protecting the company's reputation and assets. After reviewing these roles, I think you'll come to agree with me that being a project leader in today's world means a whole lot more than it used to mean (Figure 7.1).

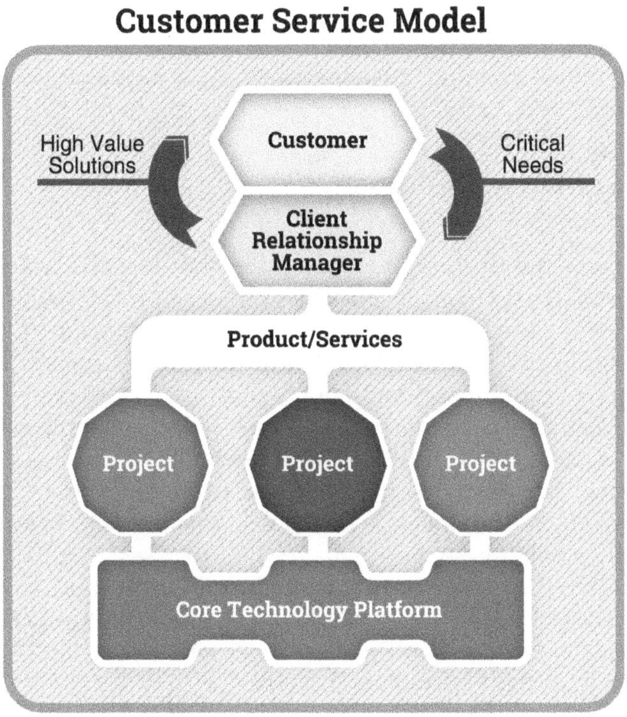

Customer Service Model

FIGURE 7.1 The customer service model.

What Is a Traditional Product-Driven Company?

Traditionally, most technology-based businesses have been product-driven, and many companies are still product-driven today. What does it mean to be driven by your product? What it means is that company management is focused on what you produce; i.e. your services or products. Product-driven companies feel that they know what their position in the world is, what it is they make or do, and then they go out and find customers to whom they can sell their products.

A classic example of a product-driven business is a high-end automobile company, a Rolls Royce, for example. If you worked for that high-end automobile company, you might have such a clear understanding of who you were and what you represent that you could sum it up by saying, "We are Rolls Royce."

What you would mean is that you have fine engineers and a management structure that is solely devoted towards making the most intelligent, beautiful, finest piece of automobile machinery in the world. Your goal would be to make the best automobile on earth. Making the fine automobiles would come first. Then you'd go out and sell them. Really, you'd hope that they would sell themselves.

For a product-driven company, the independent variable is the product and the dependent variable is the customers. Product-driven companies say to their customers, "This is what we make, would you like to but it?" To themselves, they think, "We'd better go out and find enough clients so that we can do what we do and do it well and get even better at doing it, so we can get even more clients." A product-driven company focuses mainly on its product and then goes out and finds customers.

7.4

For a product-driven company, the independent variable is the product and the dependent variable is the customers. Product-driven companies say to their customers, "This is what we make, would you like to but it?"

What Is a Market-Driven Company?

A market-driven company on the other hand does not focus first and foremost on its products. It focuses more on clients' needs and markets. Starting with its strategic plan, a market-driven, client-focused company starts with customer needs.

Rather than producing a specific product or providing a specific service, a market-driven company makes general decisions about what kind of businesses it will undertake. The company then spends much of its time and energy focusing on what the client needs. If you worked for a market-driven company, you would not have as firm an identity as the Rolls Royce employee. Rather, you would be thinking, What technologies do my clients' need? How can we help them do their jobs and run their businesses better?

In a market-driven company, the customer is the independent variable and the products are dependent. People from market-driven companies spend a lot of time with their customers. It is the customers that drive their product development and technical services.

7.5

In a market-driven company, the customer is the independent variable and the products are dependent. It is the customers that drive their product development and technical services.

It's Really a Continuum

To the uninitiated, it may seem like there's not that much difference between product-driven and market-driven companies. After all, you have to have good products to satisfy your customers, and you have to have good client relationships to sell your products.

The difference is in emphasis. It is much harder to be a market-driven, client-focused business. You have to be much more flexible. Market-driven companies have to be willing, literally, to change their internal focus, their product or project-development efforts, their project-management approaches, the types of people they hire, and the way they service their customers. Not just once—market-driven companies have to change all these fundamental aspects of their business repeatedly. They have to be willing to tailor all those parts of their business to the needs of their clients. Market-driven companies are externally focused, and that is a very difficult thing to do. Being externally focused requires disruptions to your internal organization. It requires massive change that is continuous over time.

Most Companies Think They Are Market-Driven

If you were to ask the top management of most of today's S&T-based companies, they would say that they are market-driven. Virtually, all of those managers would say that they are customer-focused. If you audited what they actually do, you'd find that most of them were mistaken. In other words, they're deceiving themselves.

You've probably heard the same argument about quality. Talking about quality has been hot for the past couple of decades. And most companies say that their focus is on quality. Some say that their most important product is quality. But if you think about it, these statements are just slogans that you almost have to say to clients. You'd be embarrassed to say, "Oh, quality isn't really that important to us."

Well right now, it's very popular to say, "We're market-driven and customer-focused." You would be embarrassed to say that your products are more important to you than your customers. The fact of the matter is that many people who say that they are customer-focused don't even know what it means.

What Is the Real Difference between Being Product-Driven and Market-Driven?

This statement sounds trite, but it's true. Being market-driven should not be a slogan—it should be a way of life. If your clients tell you that the product or service that you are now providing for them does not make them competitive in their own markets, your response should be, "We're going to have to change it." Imagine for example that your clients tell you that they like and need a service you are providing but they need it at half the cost. Your response must be to figure out a way to provide them that service at half the cost. This is not easy.

A product-driven company would tell the client that such a move was impossible. They would have to tell their customer to go someplace else and get a cheaper product. They'd explain why their own product costs what it does. They'd remind the client that they are already providing it as inexpensively as possible. The message that the product-driven company gives the client is, "I'm not going to go and change my product for you. If you don't want it, I'll go someplace else and sell it."

That product-driven company may think that it is client-focused. Possibly, they really were providing the specific product for as low a cost as possible. But that's not a client-focused company.

The market-driven company would have listened to what the client had said. They would have heard that the client liked a service but could not pay for it. The market-driven company would have already spent a lot of time with that client. Through many discussions, they would not only hear that the client liked the service, but they would know what precisely it was about the service that made it attractive and useful to the client. Knowing what the client truly liked and needed would be the first step in going back and figuring out another way to provide that service—a way that provided the client an appropriate alternative at half the cost.

Now remember, these descriptions are of the two extremes. Most companies are somewhere in the middle—there is no such thing as a purely product-driven or a purely market-driven company. What's important is where a company falls within the continuum.

Technology Companies Are Moving toward Being Market-Driven

Most successful technology-based companies realize that they must move in the direction of becoming market-driven. Simply providing technology is no longer good enough. The idea of adding value-added services on top of that technology is where most of the best companies are making their money.

Technology-based, market-driven companies operate under a customer-service model. That model has three important critical business functions. First and foremost is relationship management, second is service delivery, and third is technology development.

Relationship Management

Client-relationship management, the first important function of our customer-service model, is the heart of being market-driven and customer-oriented. Client-relationship management is the process of understanding and meeting the clients' critical needs.

Relationship management by definition means spending enough time with a select group of clients to develop a relationship. A relationship means that you get to know someone fairly well, in some form or fashion. In this case, it happens to be a business relationship, whereby you get to know and understand what the client needs to make that client successful.

The customer-service model starts with relationship management. Relationship management is the process by which a company and its clients become partners. By managing relationships with his clients or other parts of his organization, a technical professional becomes an ambassador to his own firm or his department. The relationship manager gets to know his or her clients' needs so well that he or she actually can represent them to his senior management and throughout the company.

Service Delivery

The second business function in the customer-service mode is the concept of service delivery. Service delivery is the process of bringing solutions to your clients' critical needs. Bringing solutions can mean writing a proposal, setting up a project, executing a project, or delivering a final product. The product can be producing a piece of hardware or software, running a meeting, writing a report, giving advice, or helping them through a tough negotiation with a regulatory agency. Whatever the end product, the process of setting up the project, running the project, and delivering the solution comprises service delivery.

Service delivery includes what we think of as the more traditional parts of project management, such as providing work products on time, on budget, and as promised to the client. That's what traditionally has been taught as the essence of project management. On time, on budget, service delivery. The difference under the customer-service model is that, more often than not, what you deliver is focused on customer needs, not on your own idea of what makes a good deliverable.

Technology Development

Last but not least, the third critical business function of the customer service model is technology development. Faster and faster, technology is changing and developing all the time. For any technology-based company to persist in today's climate, it must have some mechanism of developing and introducing new technology. Otherwise, the firm is not an S&T-based company. In fact, it probably is no longer a company at all.

S&T-based firms must stay on the leading edge. Stephen R. Covey in his bestseller, *The 7 Habits of Highly Effective People* [11] tells us how people—and companies—have to renew themselves. He calls this renewal "sharpening the saw." There comes a point of time when a company can do everything right—plan, organize, execute, deliver—but the saw still gets dull. You have to spend some time sharpening the saw.

If you work for an S&T-based business and all you're doing is relationship management and service delivery, over time, your technology is going to get stale. If you follow Covey, the technology part of your saw will get dull. There has to be a technology development part to your business. And presumably, the better you develop new technology, the better your service delivery will be. Service delivery depends upon technology development.

Implementing the Client-Service Model

A good S&T-based company needs to make sure it has a good client relationship-management (CRM) process, a good service-delivery process, and a good technology-development process. Sometimes one person can carry out all these functions, and sometimes a company has separate people to carry out each function. The important message is that a successful company must have an active, well-defined process and well-trained leaders in each of these areas.

For the most part, the concepts of the customer-service model are not new. They have simply been repackaged. For instance, since the beginning of time, it's been important to stay close to your customers. But one important new aspect is that good companies are developing formal procedures to make sure that they actually do what they say they are going to do.

Good companies must have well-established policies, procedures, and processes for all three of the critical business functions. They must understand who leads each of the three efforts and how those people get trained. They must know where leadership comes from and how people are evaluated. Obviously, the way to evaluate a relationship manager, for instance, must be very different from the way to evaluate a service-delivery manager. A good company must have procedures for evaluating both.

The successful S&T-based business must have a business development, marketing function, whose focus primarily is to develop strong, long-term relationships with its clients. Think about it. You cannot be client-focused unless you are willing to spend time with your clients. And this focus must begin with top management. Top managers, key people in the organization, must be willing to spend more time with the clients than they do internally.

Now, not all relationship managers are project managers. Sometimes, relationship managers are actually at a level higher. The more senior the project manager, the more he or she tends to be responsible for relationship management. And while not all relationship managers are project managers, most relationship managers know about the projects that are going on or have some oversight over them. Relationship management can be an umbrella position, overseeing multiple projects all for the same client. But if you are a very senior project manager you should be a relationship manager as well.

How does a good company evaluate the successful relationship manager? These processes are only beginning to be worked out. There is only now starting to be a formal understanding of the procedures and practices in good client-relationship management and client relationship processes (CRM). Starting in 2009, Mark Benioff has built an 8 billion dollar business, Salesforce, based on CRM software programs. My experience is that even the best technology companies are only beginning to standardize these practices.

As for service delivery, traditionally, this area is the one that companies already know best. This area is the one in which they are also best at evaluating themselves. "Faster, better, cheaper" has been the mantra of the past decade. Now the key for the successful firm is to make sure that "faster, better, cheaper" also provides the client with what he or she needs.

Finally, many S&T-based firms probably think that technology development is what they do best. But in the past, internal research and development (IR&D) was a filter-up wish list from technical staff. Overall, the entire budget probably was considerable, but it was divided up into many small chunks, spread throughout the departments, and it had very little impact on client needs. Sometimes, IR&D had very little market value. It did nothing to help clients do their jobs faster, better, cheaper, or more effectively. Now, the good companies spend the same money, but they spend it more wisely.

Today's best companies usually pick a few technology platforms, areas in which they can excel. They don't pick specific products, but they do pick general areas. The company's client-relationship managers can take a look at these platforms and what's coming out of them. The relationship managers can then work with the customers in terms of how the technology platforms can best meet customer needs.

The Role and Responsibility of the Project Leader in Executing the Customer Service Model

Now that we have a better picture of a technology-based organization and the customer service model, we can better describe the role of the project leader and his responsibilities. As with Personal and Executive Leadership, Project Leadership requires executing the fundamentals of the Performance Trilogy, Strategy, Execution, and Leadership development. Unlike the first-line supervisor, the project leader must deal with many more stakeholders as shown in Figure 7.2.

A good project leader must have a strategy for every aspect of the client-service model, including relationship management, service delivery, and technology development.

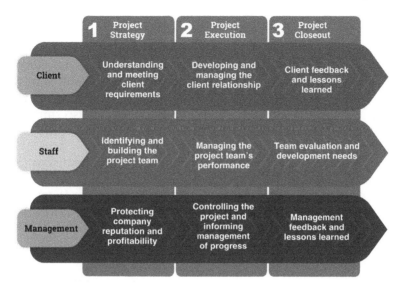

FIGURE 7.2 Stakeholder expectations.

A client strategy includes an understanding of client needs and his company's process for relationship management for key clients so as to become a good will ambassador. Executing the client strategy involves not only delivering the agreed upon product or service but also getting information and feedback from clients and further developing the client relationship. Finally, during and after completion of the project, lessons-learned meetings are conducted to learn how to improve and better serve the client in the future.

A staff strategy includes selecting talented and motivated team members and building enthusiasm for the project. During execution, staff performance is monitored and evaluated based on meeting and exceeding client expectations. During and at the completion of the project, the project leader conducts lessons-learned meetings to congratulate the team on successful project completion and identify areas for future improvement and the project and individual level.

A management strategy should include a commitment to meeting the financial commitments of the project, developing new technology, and building a long-term relationship with the client that enhances the company's reputation. Execution should include frequent project updates and communication of progress as well as problems encountered. During and after completion of the project, debriefings are held to discuss lessons learned and receive management feedback.

Since technology development is often developed on project, the project leader must be actively involved in ensuring that ideas for new technology development are discussed at project meetings. Oftentimes, new technology is developed by its IR&D department. The project manager is in a good position to give feedback to the IR&D group as to what kind of products they ought to be developing. Passing on critical client needs to the IR&D department is especially important because when working on internal projects, it can be easy to forget that the ultimate goal is to meet client needs. The project leader is in a perfect position to advise people conducting IR&D efforts. When an internal project could benefit a client, the project manager must be ready to say, "You know if you did such-and-such, my client would love that. Can you add that to your IR&D project?"

Project Leader's Role in Strategy

Vision, a key element of strategy, is somewhat an overused word these days, and it is too bad because it is such a powerful word. The skepticism probably stems from the first few definitions in Webster which describes vision as something seen in a dream or dealing with the supernatural. Most hard-nosed professionals shy away from such concepts. The more appropriate definition is further down the list: unusual discernment or foresight, the act or power of imagination. It is with this meaning that vision has meaning in our business context.

The project leader is the one person who is in the middle of multiple stakeholders, each of which only has a narrow view of the project. By judicious questioning, he is capable of developing a strategy that puts the project in a broader context. For the client, how does this project relate to his client's business success? What has proceeded this project that might affect its outcome? What is the natural follow-on to this project for his client if this project is successful? If it fails? Should these elements be factored into the project earlier?"

> **7.6**
>
> The project leader must develop the strategy as he is the one person who is in the middle of multiple stakeholders, each of which only has a narrow view of the ultimate success of the project.

For the project staff, does this project mean anything to the project staff other than employment? Does the staff have the big picture and how their specific piece contributes to the overall project mission? How important this project is to the client's success, to the company? What professional opportunities can result for staff working on this project?

For management, does the management of the company understand the value of this client and this specific project to its mission and goals? If successful, what does it mean for future work for your division? Other divisions? What would a public testimonial mean to your marketing department in developing new business? By putting all of these elements into a coherent picture, you can develop an excellent vision for your project that can be articulated to the appropriate stakeholders.

Project Leader's Role in Execution

This is where the project leader puts good intentions into action. The best laid plans are just that, plans. The world is full of ideas that have never gotten off the drawing board. How often have you heard, "aw, that's not new, I thought of that years ago"? A vision and strategy in and of itself is quite useless unless it is followed up by an initiative. Every vision has a number of specific actions that must be planned and executed to be realized. The project leader must make it happen! Whether you take it on yourself or delegate it, there must be a conscious effort to include it on your list of deliverables to meet.

An example of staff development on project will illustrate my point. You have an assistant project manager (task manager) that is ready to step up to project management. Your vision is that it is good for the staff member's career, your client will appreciate the added depth of your project management, and it will free you up for additional project work or client development. This vision is only good if you act on it. This will require initiative on your part above and beyond the day-to-day running of the project.

You will have to plan on having this task manager at more client meetings, and if that involves extra budget, you will have to spend time justifying it to your manager. It may require some training in presentation skills, which could be handled first within your project team, a safe environment, prior to presenting to the client. It may involve you delegating major project responsibilities to the task manager with close supervision for a while. You get the point. To implement your vision, you have to set specific action-oriented goals and execute them. Goals are nothing more than elements of strategy with a deadline.

Invariably, once started, your initiatives will meet with opposition and road-blocks of infinite variety. The vast majority of project leaders will be able to competently develop a vision. Some will be more inspirational than others, but most will be quite good. A significantly lower percentage of project managers will actually develop and implement specific goals and action items to achieve their vision.

Only the very best will have the courage, energy level, and commitment to work through and overcome all of the obstacles that get in the way of successfully achieving the vision. These are project leaders that are in great demand by clients and company managers, who generally reach the top of their profession. Managing change is extremely difficult. To be successful, the consequences of the change (i.e., the eventual benefits that are derived from the change) must be considerably advantageous for most people to take the risk.

Although Machiavelli's quote is over 500 years old, it is still relevant today. Its time-lessness deals with human nature. "There is nothing more difficult to take in hand, more perilous to conduct, or more uncertain in its success, than to take the lead in the introduction of a new order of things." Overcoming obstacles requires discernment, per-suasion, and plain hard work. Judgment is necessary to ensure that you can distinguish between the very difficult and the impossible. Tilting at windmills might be courageous but not very productive. The expression "choose your battles wisely" is appropriate here. Having said this, it is also important to point out that the bigger the obstacles to overcome (you are now in the vicinity of the windmill), the bigger the impact you are likely to have. Having chosen to persist, your powers of persuasion are key; the more clearly you articulate your vision and the fact-based advantages of your initiative, the greater your chances of success. There is no substitute for hard work. There are usually mul-tiple people involved, which require endless round robin discussions.

7.7

There is nothing more difficult to take in hand, more perilous to conduct, or more uncertain in its success, than to take the lead in the introduction of a new order of things.

I will give an example here that illustrates the type of challenge that the project leader faces in trying to balance the needs of multiple stakeholders. A client of mine called and requested that we accelerate a project we were working on by 3 weeks, as he needed the results to present to his board. Accelerating a project by 3 weeks would require a dramatic increase in the work effort of my team and a considerable increase in cost to the project. The challenge was how to please the client, my staff, and my management? Something had to give. So, I told my client that I didn't want to commit to the new deadline until I talked with my staff. I needed to be sure that my staff were capable and willing to do it and what the cost implications would be. I didn't want to make any promises I couldn't keep.

I discussed it with my staff and worked out a plan whereby they could compensate some of the time for their own use and worked out the cost implications with my man-ager. I got back to the client, told him we were up for the challenge, and gave him the details of what it would take including additional costs. My client was not only satisfied with my new proposal but told me how impressed he was that I wouldn't commit my staff to the new level of effort until I got a chance to discuss it with them. While it's not always easy to please all three stakeholders, a good project leader will always try.

Project Leader's Role in Development

It's a theory that's not currently prevalent in today's literature, but I truly believe that the single most important asset in a company is a talented and motivated professional staff. Twenty or thirty years ago, this theory was predominant. Many firms had slogans that echoed, "our people are our most important product." In today's world of downsizing, outsourcing, lean-and-mean staffing, and use of consultants, the theory seems to have gone out of the window. Today, it is popular to state that the customer is being the most important part of a business. Employees are treated as though they are interchangeable pawns. This is faulty thinking.

As we have stated previously, in any professionally oriented S&T-based company, the people are the product. Firms compete on the basis of whether their people are better, more motivated, or hard working than their competitors. Often, the basis of today's competition has very little to do with the products or services that a company offers. The real product is the intelligence the hard work that the company's professional staff brings to client relationships. Your staff is what makes the difference in most competitive procurements. Since the quality of your staff is the basis of competition, having competent, motivated, and trained staff provide you with an automatic leg up on your competition.

Attracting, maintaining, and training a highly motivated professional staff is the single most difficult thing to do in today's climate. It's a tough climate—business volume is unpredictable and it's highly competitive. Taking on a new staff member and then nurturing him or her is fraught with risks. But despite those risks, creating an environment where professional people want to come, where they are convinced that they will continue to develop as professionals, and where management believes that professional development is in the best interest of the company, is an extraordinary competitive advantage.

What most people want out of working for an S&T-based firm is professionally challenging work and good relationships with their bosses and their peers. Jeanne Meister has listed several rules for maintaining staff motivation [13]:

Provide clear goals

Give prompt feedback

Reward performance quickly

Treat staff like winners

Involve staff in decision making

Seek staff opinions often

Provide autonomy in work

Hold staff accountable for results

Tolerate impatience

Provide various work opportunities

Keep staff aware of upcoming, challenging goals

Notice that Meister does not include paying high salaries as a method for motivating staff. Don't be mistaken. Salaries are important. Low salaries, especially if they are out of line with the rest of the company or with the going rate for your business, can discourage the staff member who feels that his or her salary is too low. Salaries can be too low. But high salaries are not motivators. Good salaries do not make for employee satisfaction. Professionally challenging work and good personal relationships are what make people happy.

Usually, people are recruited because of their knowledge. In hiring a new staff member, managers typically look at degrees and experience. They look at what the applicant knows. In reality, when people are assigned to clients, the true value is less in what they know than in what they can do. Talent and skills are more important than knowledge. In business, those skills are very different from what the technical person has been trained to think about. Skills such as interviewing clients effectively, winning the trust and confidence of clients, and being persuasive in oral and written presentations are not what most technical people have been taught in graduate school. Diagnosing needs of clients and making judgments about those needs—these are skills that best can be developed through experience and practice.

Almost every company has staff development funds. Typically, these funds have been parceled out to various in-house and external courses. A company might have, for example, internal courses on leadership, project management, marketing, proposal writing, and project management. Over several years, every mid-level professional staff member might have an opportunity to take those courses. Some employees also might have an opportunity to take a specialized course outside the firm.

But often, work life is too hectic to take the time for courses. Project deadlines take precedence over training. Staff development funds often go underspent over the course of a year, because no one has time to spend them. And it's not just deadlines. For top management looking at the bottom line, any time that midlevel staff members are taken away from work is time that isn't being billed to a client or spent on a company project. S&T-based companies typically count on their midlevel professional staff spending upwards of 80% or more of their time on projects. Even if they have available training funds, these companies just can't afford to have midlevel professionals doing anything but spending their time to projects.

Many project managers dislike giving tasks to junior personnel. Their reasons are simple and mimic the client's reluctance to hire a new firm. Project managers typically feel that they cannot trust a junior person to do a good job. Ask those same senior project managers how they reached their own positions, and typically they will tell you about a mentor, a person who took them under his or her wing.

The projects that junior staff members work on determine the kind of professionals that they become. We all as professionals are the people we are today because of the people we have worked with and the projects we have worked on. We've taken a little from each of our experiences and learned from them.

Economic realities dictate that, in today's world, technical staff members must spend almost all their time working on projects. So, if professional development is going to take place, it almost certainly is going to have to take place on project. In fact, apprenticeships are best conducted through project work. An apprenticeship begins when a junior staff member is placed on a project team that is being overseen by a senior project leader. Once the junior staff member is on the project team, it is not so important for the project leader to teach him or her exactly what to do. It is more important to teach how to do it and why. Many senior project leaders, if they give it any thought at all, don't think that junior staff members need to know why things should be done on a project. But knowledge and skills grow enormously when people understand why they are doing something. Teaching exactly how to do something is teaching a technician's skill. To become a senior staff member, a person has to know why.

The best apprenticeships should get people involved in the planning of the work, understanding how the work fits into the whole, interpreting the work, and sharing discoveries made as the project unfolds. People should know how those discoveries might change the course of the project. Junior staff members should be trained not just how to do, but how to think throughout the assignment.

For instance, junior staff members should be trained how to write persuasive reports based on the data. Typically, clients do not want academic reports on their projects. They want findings to put into a context that will be meaningful and useful to their own stakeholders. Junior staff members must be taught to write reports that have their findings leap out to the client—no client should have to painstakingly search a report to determine the important messages.

So, a company must develop a process whereby training and staff development occurs on project. Developing such a process means that a long-term view of success must overcome the short-term concerns such as budgets, not being able to support staff development, schedules being too tight to accommodate training, or simple fear of the risks of failure.

One idea is for companies to take their staff-development funds and give them to project leaders. The project leaders could use the funds to develop apprenticeship activities on their projects. If staff-development funds go unspent and the most effective training actually takes place within projects, why not allocate the money where it will actually be spent and where it can be most effective?

For companies that prepare project plans, one of the best ways of making sure that training is conducted is to incorporate training plans into the project plans. In addition to the traditional parts of the project plan such as scope, schedule, and budget, there should be a separate section on staff development.

This mechanism forces companies to assess what opportunities are available for staff development on each and every project. For some projects, there may be no opportunities. But if there are, you can assess which members of the project team can best benefit from learning those skills. If it is explicitly written in the project plan and is part of the project review at the end of the project, it's more likely that the training will get done.

Today's companies also must develop mechanisms for rewarding senior staff who effectively develop junior staff members. Most companies today do not have any such mechanism. They don't even notice that it happens. Staff development simply is not as visible as bringing in a major project or receiving a client accolade.

Today more than ever, companies are focused on their client needs. It doesn't seem as rewarding to get involved as a coach. So, in a good company, there must be a clear message from top management that staff development is important. Good companies must spend resources on staff development and reward people who do it right.

Where the Project Leader Fits In

In my model, the function of staff development clearly falls on the project leader. This idea is very new, and you won't find it in the other books you read. Typically, the role of project manager is to bring projects on time, on budget, and meeting performance standards. Then they move onto the next one.

> **7.8**
>
> Project management is the single most important business function in today's technology-based businesses.

But we're saying, get it in on time, on budget, do a good job, really get to know the client, become an ambassador to your client's firm, and by the way, we're also expecting you to be a coach. You are a very senior person, and you have tremendous influence on the people who work here. Whether you like it or not, you are a leader who affects morale, esprit de corps, and mental energy. So, it's your responsibility to develop people. Staff development takes place on project—so staff development falls squarely on the back of the project leader.

Like I said, stretching staff generally has benefits to the client, the staff member, and the company management. It is most stressful to the project leader. The project leader assumes the risk of failure. It's another burden that we are placing on the project leader's back. All the more reason to choose your project leaders carefully and reward them handsomely both extrinsically and intrinsically as they are at the center of two of the most critical goals of your organization: satisfying clients and developing your staff.

You can now begin to see why I say that project management is the single most important business function in today's technology-based business.

☐ Chapter Summary

Project leadership is the most important business process being performed in today's organizations.

- Main interface with internal or external clients

- Steward of company reputation and profit

- Developer of organization's staff

Technology-based business—one that has products and/or services that require significant technical input and advice to ensure client satisfaction.

- Business development—Dependent on the project leader to deliver service plus product

- Staff development—Needs to be executed on project

- Business management—Must involve all staff

Customer service model—Market-driven, customer-focused businesses understand in depth that it is the customers and not the products that allow their companies to exist and flourish.

Today's best companies have found that "value-added" services are necessary if they want to stay in business.

Product-Driven Companies—The independent variable is the product and the dependent variable is the customer. Product-driven companies say to their customers, "This is what we make, would you like to but it?"

Market-Driven Companies—The customer is the independent variable and the product is the dependent variable. It is the customers that drive their product development and technical services.

Client-Relationship Management—The process of understanding and meeting the clients' critical needs.

Service Delivery—The process of bringing solutions to your clients' critical needs.

Technology Development—Mechanism for developing and introducing new technology.

Project Leader's Role

Develop the Strategy—Accounting for multiple stakeholders: client, staff, management

Manage the Execution—Maximize the project contribution to each stakeholder and manage conflicts

Coach the Development—Develop staff on project

☐ References

1. Kerzner, Harold, *Project Management, A Systems Approach to Planning, Scheduling and Controlling*, John Wiley, New York, 2006.
2. Labovitz, George, Rosansky, Victor, *The Power of Alignment: How Great Companies Stay Centered and Accomplish Extraordinary Things*, John Wiley, New York, 1997.
3. *Advameg's Encyclopedia for Small Business*, Atlantic Publishing Group, Ocala, FL, 2011.
4. Hammer, Michael, Champy, James, *Reengineering the Corporation*, HarperCollins, New York, 2003.
5. Hammer, Michael, Hershman, Lisa, *Faster Cheaper, Better*, Hammer, New York, 2010.
6. Deming, William E., *The Essential Deming, Leadership Principles from the Father of Quality*, McGraw Hill, New York, 2013.
7. Goetsch, David, Davis, Stanley, *Quality Management for Organizational Excellence*, Pearson Education, New York, 2014.
8. Beckwith, Harry, *Selling the Invisible*, Warner Books, New York, 1997.
9. Moore, Geoffrey, *Crossing the Chasm*, Harper Collins, New York, 2013.
10. Stack, Jack, *The Great Game of Business*, Currency Books, New York, 1992.
11. Covey, Stephen R., *The Seven Habits of Highly Effective People*, Fireside (Simon and Shuster), New York, 1989.
12. Dychè, Jill, *The CRM Handbook, A Business Guide to Customer Relationship Management*, Addison Wesley, Boston, MA, 2002.
13. Meister, Jeanne, Willyerd, Karrie, *The 2020 Workplace, How Innovative Companies Attract, Develop, and Keep Tomorrow's Employees Today*, HarperCollins, New York, 2010.

Executive Leadership
Leading Organizations

Leading S&T-Based Organizations
Seven Critical Business Processes

Organizations that rely on science and technology (S&T) present unique leadership challenges. I have defined leadership in terms of the Performance Trilogy®, consisting of three distinct roles: developing and communicating a winning strategy (leading); flawlessly executing the strategy (management); and ensuring the performance and development of the staff responsible for doing the work (coaching).

Let's address these three roles one at a time and discuss why leading S&T organizations is so unique. When developing a strategy in most organizations, innovation is a highly desirable element but not essential for its success. For many companies, innovation is broadly viewed to include new business models, marketing approaches, and delivery systems. In S&T-based organizations, the main focus is on technical innovation. Depending on the industry, research and development (R&D) budgets can range from 0% to 3% of revenues. In S&T-based organizations, however, (e.g., biopharma, medical products, information technology) technical innovation is a matter of survival, and R&D budgets are often 5%–10% of revenues. Indeed, there are over 2,000 public and private global research institutions whose entire business is R&D. Eighty percent to one hundred percent of their revenues are derived from conducting R&D and providing advice to their government and industrial clients based on sound science. Combine this with the fact that new product life cycles for S&T-based companies have been reduced from 7–10 years to 18 months in many cases, and it is easy to understand the difficulty in developing a sustainable winning strategy based on technology.

In executing their strategy, S&T-based organizations utilize similar business processes widely taught in business schools. However, the application of some of these processes is once again fairly unique, particularly managing the R&D process. The R&D process involves creating organizational value through research, development, and exploitation of S&T that increases revenues, reduces costs, and minimizes risks. Leaders of S&T organizations stand at the nexus of two cultures, business and science, and need to have the skills necessary to bridge this wide gap in language and perspective (Figure 8.1).

FIGURE 8.1 Bridging the gap.

In this chapter, seven critical business processes that need to be mastered to successfully manage an S&T organization are identified.

The third phase of the Performance Trilogy may be the hardest of all to lead, ensuring the performance and development of scientists and engineers executing each of the critical business processes and doing the work. I have often been quoted as saying that "If you can lead scientists and engineers you can lead anyone." I mean this as a compliment. Scientists and engineers by their very nature and training are taught to be skeptical. The peer review process demands that every new idea or concept be thoroughly examined and critically challenged before acceptance. Convincing scientists and engineers requires strong evidence that your strategy is desirable, achievable, and beneficial to them. This oftentimes leads S&T managers to avoid involving those scientists in the strategic planning process, who are the most likely to improve the strategy by their tough questions and technical input. A good book on the culture of scientific organizations worth reading is Lab Dynamics [1].

For those scientists and engineers who have accepted the strategy in theory, translating the strategy into organization-driven performance and financial goals that align with their personal aspirations represents an additional leadership challenge. Senior technical professionals are highly intelligent and independent minded and oftentimes have a work agenda that is inconsistent with the organizations. This is where coaching plays an important role in understanding the talents and motivation of each individual. Only then can you "cast" them in the right roles that balance organizational and personal needs.

In any S&T-based organization, there are seven critical business processes that must be executed flawlessly (i.e., managed) to ensure high performance and achievement of the organizations' vision. The Performance Trilogy is a useful tool for these processes, as they each involve leading the development and communication of a strategy; managing the execution of that strategy; and coaching independent-minded staff to perform.

8.1

The R&D process involves creating organizational value through research, development, and exploitation of S&T that increases revenues, reduces costs, and minimizes risks.

Critical Business Processes

FIGURE 8.2 The seven critical business processes.

The seven critical business processes that require superior S&T leadership to achieve success are shown in Figure 8.2. This business model is directly applicable to S&T organizations performing contract research and technical consulting for clients. However, it is also useful for internal R&D departments that provide the same service for internal clients within their company.

Strategic planning, R&D management, and financial management are the first three vertical processes that involve developing and communicating the strategy up and down the organization and executing that strategy to achieve business and financial goals. While senior and middle management traditionally perform this function, in R&D organizations, senior technical staff play an important role in providing technical input. Business development, project management, and product development are the next three horizontal, cross-functional processes designed to deliver products and services to the satisfaction of clients and stakeholders.

The final process is the overarching process of renewal through the acquisition, development, and supervision of staff throughout the organization. Organizations have differing definitions than these, so I will provide my definition of each for clarity.

☐ Strategic Planning

The strategic planning process starts with the development or updating of the organization's mission, vision, and values and finishes with a strategy designed to fulfill its mission and achieve its vision. The mission is usually a variation of developing, deploying, and exploiting S&T to create value for the organization depending on the industry sector. In the case of government research institutions, the mission is often expanded to include recommending policies and commercializing technologies based on sound science to stimulate national economic and social development.

Developing a winning strategy first requires a codified process consisting of several steps. These include a thorough understanding of the organization's strengths, weaknesses, and assets/resources available to it (self-assessment); the strengths, weaknesses, and strategy of competitors fighting for the same market space (competitor analysis); intelligence from existing clients on their problems and opportunities (client feedback); awareness of competing and emerging technologies that could disrupt the strategy (technology assessment); a thorough knowledge of market and industry trends (market analysis); a clear understanding of the regulatory requirements or barriers to market (regulatory analysis); and finally an awareness of the political and social environment that could positively or negatively affect the organization's brand (brand awareness).

Collecting and synthesizing this disparate data and information into actionable intelligence is the organizations' first major challenge requiring in-depth knowledge of both technology and business. This is why scientists or engineers lead most S&T organizations. Even in the case of startup companies led by a business entrepreneur, he/she is usually partners with a chief technology officer to oversee the technology strategy.

Armed with this information, senior management then develops a business strategy consisting of several thrusts designed to allocate its limited resources to protect and exploit its technologies; compete in a favorable market space, and recruit and retain top-level talent. The most important focus of the strategic planning process is making decisions on where to allocate scarce resources. Too many organizations get bogged down on the process writing the plan and lose sight of the real importance of the planning process, which is to make tough decisions about where to allocate resources.

8.2

Too many organizations get bogged down on the process writing the plan and lose sight of the real importance of the planning process, which is to make tough decisions about where to allocate resources.

☐ R&D Management

Developing the organizational strategy and communicating it in a compelling way is the foundational process led by senior management with input from senior technical, marketing, and financial staff. Translating and executing the broad organizational strategy at an operational level is the major role of S&T managers. There are four main process steps that every S&T manager must master: translating the organization's business strategy into an S&T strategy that is aligned with the organization's vision and mission and creates value; managing S&T performance (both technical and financial) to meet the organization's goals; managing the considerable S&T resources; and overseeing the delivery of the execution process to include business development, project management, product development, and career development.

Developing an R&D strategy and operational plan that fully supports the organization's business strategy is one of the biggest challenges facing S&T directors. In government and contract R&D laboratories, their organizations are usually divided into divisions or programs focused on specific strategic business thrusts driven by clients or public policy. For product-driven R&D organizations, the main organizational strategy is focused on technology platforms that can continually generate new products. In my experience, R&D strategies are oftentimes poorly aligned with the organization's business strategy or not specific enough to be able to monitor and manage performance.

Once the S&T strategy is developed, the performance management process is used to ensure that the strategy is flawlessly executed. This process involves setting specific performance goals for the director of R&D and cascading these goals down through division, program, department, and unit managers. The larger the organization, the more difficult this process is to obtain clear alignment up and down the chain of command. More importantly, there is often a lack of specificity in the goals that makes it very difficult to monitor and evaluate progress.

In addition to the vertical processes of strategy development and performance management to satisfy senior management, S&T directors must oversee proper execution of the horizontal, cross-functional processes led by their technical staff to satisfy clients and stakeholders. These include business development, project management, and product development. And finally, what the S&T directors must manage is the acquisition and utilization of expensive facilities and laboratory resources and the career development of their staff.

☐ Financial Management

Regardless of structure and governance, all S&T organizations, whether they be public or private, profit or nonprofit, must at a minimum generate sufficient revenues to cover their costs to remain viable. Since most S&T organizations are capital-intensive (expensive laboratory facilities and scientific equipment), this means that the organization must earn a net operating profit after taxes to at least cover its cost of capital. This is the basis of value creation, and every S&T manager must have a thorough understanding of how to create value in the investment decisions they make and their operating performance.

It is unfortunate that many S&T managers do not have any formal training in financial management, having been promoted to line management positions directly from their technical positions. Too often I have witnessed condescending attitudes from CFO's who think that business finance is too complicated for technical managers to understand. What they are implying is that having completed university courses in calculus, differential equations, and thermodynamics, technical managers can't handle arithmetic! While there is a lot of specialized professional vocabulary associated with the financial profession, once the vocabulary is learned, the mathematics boils down to simple arithmetic. I strongly urge all senior technical staff take a course in financial management to help them better communicate with and understand the financial decisions made by top management.

The first step in the financial process is the investment step. Since R&D oftentimes involves large investments in facilities, capital equipment, and product development, it is important to make sure that each investment creates value, i.e., has a sufficient return on investment. This involves making sure that the net present value of future revenue streams exceeds the cost of the investment (or in layman's terms, the future revenues back calculated into today's dollars exceed the original cost of the investment). As I said, once you learn the vocabulary of finance, the math is pretty simple. The difficult part comes in determining what to count as an investment versus the cost of doing business.

For example, if budgets for technical services and quality assurance are allocated to R&D investment as opposed to product or process support, the return on investment will be artificially low and important projects that are worth funding will not

8.3

Once you learn the vocabulary of finance, the math is pretty simple.

make the cut. An S&T manager must champion the investment in R&D and be able to back up requests with sound financial data.

The second process step in financial management is the budgeting process. There are many books written on the subject. What I want to emphasize is that the budgeting process needs to be integrally related to the organizations' S&T strategy. The definition of strategy is the allocation of scarce resources. If the organization had unlimited resources, it wouldn't need a strategy. It could just spend money on everything it could think of to achieve its goals. Budgeting is the process of allocating the limited funds available to those activities with the highest potential of achieving the organization's performance goals. So, strategy drives the budgeting process that is all about making decisions. As I have said many times before, too much time and energy are wasted on the process and filling out forms when the fundamental purpose of the exercise is to make decisions.

One last point that I would like to emphasize is the link between strategy and employee engagement. Too often, strategies and operating plans are generated without the complete understanding and buy-in by key senior scientists who are then charged with meeting performance goals that they either don't understand or agree with. This is why the third phase of the Performance Trilogy is so important, coaching staff to ensure performance and development.

☐ Business Development

The first lesson one learns in business school is that a business doesn't exist until the first sale is made and doesn't develop without customers. Nothing in business happens until a sale is made. The term business development has many different meanings in the business world. I am using the term broadly to define the comprehensive marketing and sales process from building awareness to closing the sale. The business development process is the first critical step in the execution of the organization's strategy and involves several cross-functional units, including public relations, marketing, and R&D.

Once again, there are thousands of books written on the sales process and major departments in business schools that teach it, but selling S&T is unlike any other kind of sales process and one of the most difficult to execute. Selling S&T has been excellently described in *Selling the Invisible* by Harry Beckwith [2], a book I highly recommend. Think of it this way: imagine buying a product that you're not sure exactly what you will be getting or whether or not it will work, nor can you try it out before you buy it; but you must begin paying for it right away with the promise that if it does work, it will be great. It's no wonder why it is so difficult to sell R&D projects either internally to management or externally to clients!

Organizationally, the marketing and public relations departments have the role of increasing awareness; i.e., communicating who the company is, what it is known for, what are its services, and the benefits of these services and what makes the organization different and special (key elements of branding). The R&D departments are responsible for driving the development of new business by building strong client relationships and selling technical solutions/products with support from the marketing department. The role of market intelligence (i.e., who is competing with similar products/services in the chosen market space and how good are they) is shared by both marketing and R&D.

Given the broad scope of business development activities and the interconnected roles and responsibilities of marketing and R&D departments, developing and executing a seamless business development process presents major leadership challenges. I have seen it work extremely well and very poorly. In my opinion, the key to success is not just the competence of the respective individuals, but their mindsets and attitudes toward the two professions.

The best marketing professionals I have worked with have a deep appreciation and wonder for S&T and enjoy working with and for scientists. They take the time to reach out and learn as much as they can from them and then use their own talents to communicate science in a compelling way that accentuates its value and benefits to clients. They understand the value that they bring to the business development process but never lose sight of the fact that it's not about them. The product/service is all about the scientists and their technical staff.

On the other hand, the most successful R&D scientists recognize their own limitations and value what the professional marketing staff bring to the process. They are willing to suspend their egos and be challenged by marketing staff as to why their work has value and what makes it any better than the competition. They understand that communicating what they do (i.e., selling the invisible) is extremely difficult, and the marketing staff can't do it by themselves, but it must be done together. This level of alignment is rarely achieved in S&T organizations but is essential for achieving success.

☐ Project Management

While the business development process comes first for the reasons just mentioned, the most effective business development that an organization can conduct is a job well done. Successful organizations depend on repeat business. Since buying R&D is a highly risky proposition, clients tend to minimize their risk by contracting with organizations with a track record that they know and trust. Also, new clients will depend a great deal on referrals before purchasing R&D. Most R&D is conducted through technical projects that deliver reports, recommendations, or products to clients. This is why I believe that the project management process that delivers products and services to clients is the most important of all the execution processes.

Once again, there are many academic sources and practical guides to project management [3–5], including professional societies such as the Project Management Institute that issue certifications (Project Management Book of Knowledge, PMBOK). The science of project management is well understood. The concept of work breakdown structures, Ghant charts, critical path analysis, and earned value that were developed to put a man on the moon and perfected on large engineering and construction projects is now used on most technical projects. In my experience, the major difficulties in executing a technical project lie not with the science of project planning (although difficult in and of itself), but in leading the project and satisfying the needs of multiple stakeholders with different needs and expectations as discussed in Chapter 7.

The client wants his product/service delivered on time with high quality at a competitive price. The management of the organization desires the project be conducted with distinction and minimum risk and flexibility of resources while meeting the budget and making a profit. The project staff wants clear direction, sufficient time to do their work properly, and minimum disruption or changes once the work commences. Regulatory agencies (i.e., the Environmental Protection Agency, EPA, the Food and Drug Administration, FDA) demand that the work be conducted according to mandated guidelines (e.g., good laboratory practices, GLP, good manufacturing practices, GMP) and environmental standards. There is a myriad of conflicts in trying to meet all of these needs simultaneously, and the solutions are not contained in the textbooks or manuals! A few examples given later will illustrate the point.

The first conflict oftentimes is setting expectations. During the proposal phase of a project cycle, senior marketing, and technical staff make many promises in order to "close the sale." In the worst case (and I've seen it), management signs off on the proposal that commits more than can be delivered; with staff and resources that may or not be available at the start of the project; with an unrealistic budget given the proposed work scope. The assigned project manager starts his project in a deep hole that can be very difficult to negotiate out of.

Another common problem encountered is often "scope creep." This refers to changes in the conduct of the work scope caused by the client's need for additional information or experiments that involve additional costs. The client's perspective most often is that the additional work is needed to ensure success of the project or answer particular questions raised by his management and that it should be part of the original agreed upon contract cost. Your management's perspective is that this additional cost will result in a budget overrun and loss of profit, unless a "change order" is issued asking the client for more money. Your project staff's perspective is that they are not willing to put in their own personal time without compensation. Resolving such a conflict is difficult and will challenge the best of project managers.

An additional problem is the lack of awareness of the project manager (and his management at times) to truly understand the cross-functional nature of the project management process. The project team not only consists of the project leader and his technical staff but also regulatory, human resources (HR), marketing, purchasing, financial, and billing staff. Each of these team members has an important role to play in ensuring a high-quality product/service is delivered, invoiced, and paid for on time. By focusing project communications solely on the technical staff, the project manager loses the opportunity to develop ownership in his project by the entire team and has difficulty getting their attention when their intermittent services are required.

☐ Product Development

So far, we have been describing critical business processes of an organization's existing products or services. The lifeblood, however, of an S&T-based business is product development as innovation is the major differentiator among technology organizations. It is important to distinguish between technology development and product development. R&D managers lead the technology development process to harness the creativity of research staff and shepherd innovative ideas

8.4

The primary role of an R&D manager is to create organizational value from research, and it is all about the nexus between technical expertise and market insight.

through research, development, testing, and evaluation. In my opinion, the primary role of an R&D manager is to create organizational value from research, and it is all about the nexus between technical expertise and market insight.

Product development on the other hand is a cross-functional process led by a marketing/commercialization team supported by R&D and finance. Product development builds on and expands technology development to include product/packaging/service innovation, process innovation, business model innovation, new relationships with customers/markets/partners, and novel operations and supply chains. The primary role of the product development manager is determining whether the envisioned product/service will sell and it's all about market insights.

The staffs of these two processes represent very different cultures. Technology development staff represent homogeneous technical teams led by a research leader that deal creatively with uncertainty. The result of their work is often fuzzy, unpredictable, and nonlinear with an end point that identifies technical feasibility or proof of concept. Product development on the other hand involves cross functional teams led by a business leader that are focused and disciplined with a sense of urgency and an endpoint of a well-defined product and customer set.

While innovation is the Holy Grail for most technology organizations and much emphasis is being placed on it, there are good reasons why it is difficult to achieve. In my opinion, innovation is more a culture than an event. The technology development process needs to be led so that creative technical staff will have the freedom to experiment, be encouraged to build an innovation pipeline of ideas, and given the latitude to fail. Trying to control or micromanage this process is a sure-fire way to stifle innovation.

The product development process, on the other hand, needs to be closely managed so that the organization can make wise investment decisions, as the costs of development increase exponentially from the technical feasibility stage to product launch. At each stage of product development, the business case must remain compelling to justify the next higher level of investment.

In my experience, managing the handoff from technology development to product development is the Achilles heel of the innovation process. I concur with most innovation surveys by business schools that point to the involvement of senior management, even the CEO, as a key success factor in innovation. The reason in my opinion is to facilitate the appropriate handoff from R&D to marketing/commercialization and ensure that the cultural transition is managed appropriately. Most importantly, since most technology initiatives are killed during the product development process, how senior management communicates the rejection plays an important role in developing the culture.

8.5

Managing the handoff from technology development to product development is the Achilles heel of the innovation process.

I would like to share a personal experience that illustrates this point. As an R&D manager, I championed the development of a new environmental technology that I felt could revolutionize the waste management field. After 2 years of technology development, I presented the proof of concept to senior management who agreed to form a commercialization team to develop the business case. I led that team for over a year through the technical market feasibility studies, prototype development and evaluation, and financial modeling and presented the business case to senior management.

In the end, the management rejected the idea as not fitting with its strategy. Needless to say, I was upset. The very next day I was called into the office of the corporate officer who made the decision. I paraphrase his words to me: "Tony, I know you must be disappointed with the organization's decision not to proceed with your idea. You worked really hard and put together a great case for the business. Please don't be discouraged, as our organization needs more leaders like you who are willing to bring us good ideas even if we can't fund them all. I encourage you to continue and look forward to your next try." As a result of that meeting, I gained valuable perspective on the importance of leadership and how it influences organizational culture.

☐ Staff Development and Renewal

Staff development and renewal is the overarching process of talent acquisition and development that underpins all of the other critical business processes. I am defining the staff development and renewal process broadly to include the acquisition, training, supervision, and career development of staff throughout the organization. The prevalent view of this process is that it is the responsibility of the HR department. Nothing could be further from the truth. HR plays an important role in organizational development by administering the performance management and career development process, counseling managers on best practices, and advising managers on individual staff decisions. However, it is the responsibility of each supervisor and manager throughout the organization to lead the process and make decisions, whether he is managing a project, program, department, division, or the entire organization. The three key elements of the organizational development and renewal process include succession planning, hiring and promoting, and coaching for development and performance.

Succession Planning

The overall process of staff development and renewal starts with succession planning. Succession planning is the foundation of organizational development that underpins the entire process. By succession planning, I mean the systematic process of selecting the right staff for each job function and having replacements ready if and when the selected staff are promoted or leave the organization. This is particularly important for leadership positions within the organization, be they technical or managerial.

> **8.6**
>
> Succession planning is the foundation of organizational development that underpins the entire process.

Take sports teams, for example, who must perform at extremely high levels of performance at all times despite injuries or defections to other teams. The best performing teams start by selecting great talent and always seem to have players to substitute when talent is lost with little loss in performance. This doesn't happen by accident. Team rosters are developed with succession planning in mind and backup players for each of the key positions.

Unfortunately, this doesn't always happen in S&T organizations. Too many times, I have seen major programs with substantial budgets build around one key technical person with no obvious backup in case that person becomes ill or leaves the organization. Vacant technical leadership and management positions can take 6 months or longer to fill, creating quality problems and causing serious delays in organizational commitments.

I call these single points of failure and have seen hundreds of them in organizations I have worked with.

A good succession plan identifies gaps in current key positions in the organization as well as backups for each of these positions. This plan then drives the next two processes designed to fill these gaps, hiring and promoting and coaching for development and performance.

Hiring and Promoting

The hiring and promotion process should be driven by the succession plan, with the ultimate goal of having talented and motivated staff in each key leadership position with two potential backups. While difficult to achieve, identifying gaps and single points of failure in the plan will focus management attention on the top staffing needs. In addition to backup candidates within your organization, a useful addition to your recruiting strategy is to identify a list of individuals outside of your organization that could fill important positions if and when they become vacant (sports teams with limited rosters call this their taxi squad).

When hiring or promoting staff, most S&T managers place most of their emphasis on qualifications and experience in their evaluation criteria. While such selection criteria are important, they are not sufficient to ensure a high success rate. I have interviewed hundreds of S&T managers in my consulting and course work who have said that they were very satisfied with only 50% of their hiring and promotion decisions. Worse still, they admitted that about one out of every three hiring and promotion decisions turned out poorly, resulting in costly damage control and hard feelings.

I have come to believe that there are good reasons for suboptimal outcomes of the interview and evaluation process and once understood, the success rate can be dramatically improved. Two key questions need to be definitively answered during the evaluation process: "Can the candidate do the job" and "will the candidate do the job?" The first question deals with qualifications and experience and receives the majority of emphasis and evaluation. For the most part, this question can be answered pretty accurately, although it is sometimes difficult with outside candidates to tell whether they have 10 years' experience or 1 year's experience ten times.

It is the second question, "will the candidate do the job," that is rarely answered very well and is the key to improving the success rate of a hiring or promotion decision. Answering this second question deals with evaluating the specific job requirements and the talent and motivation of the candidate. This information involves evaluating leadership attributes that are harder to discern but will increase the success rate of the hiring or promotion decision dramatically. Each vacant job will have specific leadership requirements depending on the degree of difficulty of the job and the magnitude of change required.

Will the job entail strong leadership skills that require changes in the status quo? Are the changes needed strategic or operational? Will strong management skills be needed to get the job done? Will coaching skills be needed to develop and motivate staff? In Chapter 4, I have described nine key attributes that need to be evaluated to help answer these questions. The biggest mistake I see S&T managers make is in assuming that these attributes represent "soft" skills that can't be measured. By delving more deeply into this area, "hard" data can be extracted, which will result in much better hiring and promotion decisions.

Coaching for Development and Performance

The development and renewal process actually begins once a comprehensive succession plan has been put in place and positions are filled by hiring and promoting the best available staff. This is a dynamic process with much flux that requires

8.7

An R&D organization left to its own devises will tend toward mediocrity.

constant vigilance and intervention. I believe that an S&T organization left to its own devises will tend toward mediocrity. This is because outstanding staff are highly mobile, and a certain percentage of them will leave the organization over time for better career opportunities. On the other hand, mediocre staff rarely leave organizations where they have comfortable positions unless forced to. Given the nature of S&T, the competition for talented technical staff is often the key factor in organizational success, making talent development and renewal a critical business process.

Excellent coaches first focus on staff development before discussing performance goals. In coaching for development, the focus is on how the organization can meet the needs and aspirations of the employee. In most performance planning processes, staff development is an afterthought and rarely given much attention. I believe that this is one of the main reasons why organizations fail to reach their goals. In my opinion, the first priority in performance planning should not be determining performance objectives, but on the development process, keeping in mind the gaps identified in the succession plan. High performance is achieved when staff are self-motivated and give 110% to achieving organizational goals. Staff are motivated by tailored performance goals that match and challenge their skills. It is nearly impossible to develop tailored performance goals, unless sufficient time is spent getting to know the strengths, weakness, aspirations, and motivations of staff. This can only be accomplished by starting with the needs and aspirations of the staff member. More detail on performance management and coaching will be presented in Chapters 10 and 11.

☐ Chapter Summary

In executing their strategy, science-based organizations utilize similar business processes widely taught in business schools. However, managing the R&D process is fairly unique.

The R&D process involves creating organizational value through research, development, and exploitation of S&T that increases revenues, reduces costs, and minimizes risks.

There are seven critical business processes in science-based organizations.

Strategic Planning—Includes a self-assessment; competitor analysis; client feedback; technology assessment; market analysis; regulatory analysis; and brand awareness.

R&D Management—Includes translating the organizations business strategy into an R&D strategy that is aligned with the organization's vision and mission and creates value; managing R&D performance (both technical and financial) to meet the organization's goals; managing the considerable R&D resources; and overseeing the delivery of the execution process to include business development, project management, product development, and career development.

Financial Management—Includes a thorough understanding of how to create value in investment decisions and budgeting to ensure operating performance.

Business Development—Marketing and Public Relations departments increase awareness; i.e., communicating who the company is, what it is known for, what are its services, the benefits of these services, and what makes the organization different and special (key elements of branding).

The R&D departments drive the development of new business by building strong client relationships and selling technical solutions/products with support from the marketing department.

The role of market intelligence (i.e., who is competing with similar products/services in the chosen market space and how good are they) is shared by both marketing and R&D.

Project Management—The major difficulties in executing a technical project lie not with the science of project planning, but in leading the project and satisfying the needs of multiple stakeholders with different needs and expectations.

Product Development—A cross-functional process led by a marketing/commercialization team supported by R&D and finance. Product development builds on and expands technology development to include product/packaging/service innovation, process innovation, business model innovation, new relationships with customers/markets/partners, and novel operations and supply chains.

Staff Development and Renewal—Succession planning is the systematic process of selecting the right staff for each job function and having replacements ready if and when the selected staff are promoted or leave the organization and is the foundation of staff development that underpins the entire process.

The hiring and promotion process should be driven by the succession plan with the ultimate goal of having talented and motivated staff in each key leadership position with two potential backups.

Coaching for Development—The first priority in performance planning should not be determining performance objectives but on the development process, keeping in mind the gaps identified in the succession plan.

☐ References

1. Cohen, Carl, Cohen, Suzanne, *Lab Dynamics*, Cold Spring Harbor Press, Cold Spring Harbor, NY, 2012.
2. Beckwith, Harry, *Selling the Invisible—A Field Guide to Modern Marketing*, Warner Books, New York, 1997.

3. Kerzner, Harold, *Project Management, A Systems Approach to Planning, Scheduling and Controlling*, John Wiley, Hoboken, NJ, 2006.

4. Westland, Jason, *The Project Management Life Cycle*, Kogan Page, London, 2006.

5. Dinsmore, Paul C., *Human Factors in Project Management*, AMACOM, New York, 1990.

Leading the Strategy
Setting the Direction and Developing the Road Map

In the previous chapter, I presented an overview of the critical business processes that need to be mastered to successfully manage a science and technology (S&T)-based organization. The first of those processes is strategic planning, which establishes the foundation and provides direction to the organization. Indeed, it is the first pillar of the Performance Trilogy®. While developing a business strategy in general is a difficult process, a winning S&T strategy is infinitely more challenging [1].

☐ Strategic Planning Process

The strategic planning process starts with developing or updating the organization's mission, vision, and values and finishes with a strategy designed to fulfill its mission and achieve its vision. The mission is usually a variation of developing, deploying, and exploiting S&T to create value for the organization depending on the industry sector. In the case of governments and research institutions, the mission is often expanded to include recommending policies and commercializing technologies based on sound science to stimulate economic and social development.

Developing a winning strategy first requires a codified process to acquire the relevant data and information upon which to make decisions. Most organizations have developed variations of a SWOT analysis (Strengths, Weaknesses, Opportunities, and Threats) to generate this information. While somewhat useful, it is not comprehensive enough to provide all the critical information needed. A more robust process consists of seven steps shown in Figure 9.1. Steps 1–6 involve the collection of data and information, and Step 7 represents a synthesis of this information, resulting in setting the direction of the organization, developing a road map, and making decisions about where to deploy organizational resources.

Strategy Synthesis

FIGURE 9.1 The strategic planning process.

Developing a winning strategy is the organization's first major challenge requiring an in-depth knowledge of both technology and business. This is why scientists or engineers lead most science-based organizations. Even in the case of startup companies led by a business entrepreneur, he/she usually partners with a chief technology officer to oversee the technology strategy.

Armed with this information, senior management then develops a business strategy consisting of several thrusts designed to allocate its limited resources to protect and exploit its technologies; compete in a favorable market space; and recruit and retain top-level talent. The most important focus of the strategic planning process is making decisions on where to allocate scarce resources. Too many organizations get bogged down on the process and writing the plan and lose sight of the real importance of the planning process, which is to make tough decisions about where to allocate resources.

9.1

Too many organizations get bogged down on the process and writing the plan and lose sight of the real importance of the planning process, which is to make tough decisions about where to allocate resources.

There are two very important considerations to take into account that are critical to successful strategic planning: the shortening of the planning cycle and the continuous nature of data collection.

As previously mentioned, the cycle for new product/service development now ranges from 12 to 18 months. This oftentimes makes elements of traditional 5-year strategic plans obsolete during the proposed planning cycle. To counteract this, science-based organizations are reviewing and updating their 5-year plans midterm and making appropriate adjustments based on new information. To take this a step further, strategic thrusts and institutional objectives in support of the strategic plan should be updated annually and strategic objectives reviewed on a quarterly basis. In this way, the organization's strategy can be refreshed in line with new technical developments and marketplace changes. This will be discussed in more detail in the next chapter on managing the execution (Figure 9.2).

The second consideration is to seriously consider the development of an institutional Knowledge Management System. This is particularly important for large organizations where new information on technology and market trends and competitor analysis are happening continuously and collected by a broad spectrum of staff. Without a systematic way of acquiring and storing such information and retrieving it for strategic planning purposes, leads to the first reason I have mentioned for failed strategies, organizational ignorance. It's not that the information isn't located someway in the organization, but it is not accessible to those doing the planning and is therefore ignored.

While a discussion of Knowledge Management System is outside of the scope of this book, my advice is to keep it simple and embed it into the everyday processes that are being managed. For example, the questions raised in this chapter that need to be answered for strategic planning purposes should be built into trip reports on technical conferences, minutes on client review meetings and client feedback reports, lessons-learned reports at the conclusion of projects, minutes to strategy meetings, etc. Knowledge management becomes effective when it is embedded into the culture and through everyday reports of your organizational processes. It then becomes easier during the strategic

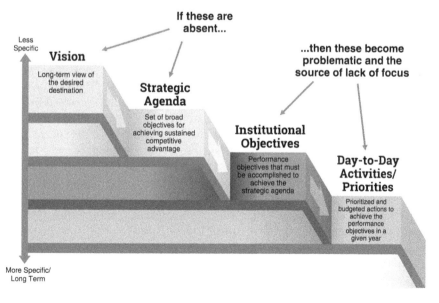

FIGURE 9.2 Linking strategy to results.

planning process to retrieve valuable information collected over previous months and years through key word searches.

☐ STEP 1. Self-Assessment—A Thorough Understanding of the Organization's Strengths, Weaknesses, and Assets/Resources Available to It

The first step in the strategic planning process is conducting a self-assessment. The purpose of the self-assessment is to gain a thorough understanding of the organizations' strengths, weaknesses, and assets/resources available to it.

On the surface, it would appear that conducting a self-assessment would be the easiest of the seven strategic planning steps to conduct. After all, all the data you need are readily accessible within your organization. It turns out that most self-assessments suffer from self-deception due to ignorance, arrogance, or both. The larger the organization, the more difficult it is to get the critical information you need, as it is often dispersed throughout the organization. Ignorance of key organizational strengths can often result in missing strategic opportunities. Ignorance of key organizational weaknesses can result in making poor decisions that adversely affect the success of the strategy during implementation.

9.2

Most self-assessments suffer from self-deception due to ignorance, arrogance, or both.

On the other hand, too often senior managers suffer from overestimating the strengths and minimizing the weaknesses of their organizations. This arrogance, often a result of being enamored with a particular technology or disenchanted with the behavior of a technical team despite the value of their outputs, leads to faulty decision making.

Having been involved in leading several turnaround efforts in S&T-based organizations, I have developed a healthy skepticism for existing self-assessments. For example, one organization I was asked to turn around was losing key long-term clients and was convinced that the cause of its problem was that its costs were too high and blamed the organization's overhead rate. It turned out that the key managers had overestimated the quality of the products they were delivering to clients. When I spoke to several clients, they expressed disappointment with the quality and timeliness of deliverables. This led to a key element of a strategy going forward to improve quality that resulted in recapturing several of client accounts and a return to profitability.

In another example, an organization that I took over had just experienced a major setback in a key strategic initiative. When I asked several staff scientists what they thought went wrong, they all expresses the same thought. "We knew all along that the strategic initiative wouldn't work" and had convincing evidence to back up their claim. When I asked why this wasn't brought to the attention of senior management, they said that when they expressed concerns at a meeting, management reaction was so negative that they thought best to keep their mouths shut.

One of my managers whom I admire once told me that the best managers have a bit of paranoia in them, always questioning whether their products and services are good enough and whether competitors are close to outcompeting them. This is good advice when conducting a self-assessment. Err on the side of caution by challenging all of the existing assumptions.

The three key evaluations to focus on are your (i) products/services, (ii) key processes, and (iii) people; the three Ps. When evaluating your products/services, there are only two foolproof, unbeatable strategies: having a desirable product or service that no one else has or delivering your product/service at a lower cost than your competitors. While these are great strategies, they are rarely seen in today's marketplace. Most strategies fall somewhere in between these extremes and require an honest, unvarnished assessment of the value that your product or service provides versus your competition (what the business community calls your value proposition).

Product/Service

The strategies of most S&T-based organizations depend on having a superior product/or service based on the latest technology or sound scientific principles. This is a double-edged sword. The life cycle of technical products/services has been significantly shortened over the past decade. It is now not unusual to see product life cycles in the one to 2-year timeframe rather than the more traditional 7–10 years.

During data synthesis (Step 7), you will be comparing your present products/services to that of your competitors generated in Step 3, competitor analysis. Also, you will be comparing your future products/services based on Steps 4 and 5, technology trends and market analysis.

To aid in the process of self-evaluation of your products and services, it helps if you have developed several codified processes (part of your Knowledge Management System) that capture and store feedback on your products/services *as they occur*. I view these as "intelligence reports."

For example, a systematic client feedback report can capture valuable data on the quality and competitiveness of your product versus your competition. A periodic client feedback report can provide you with continuous data on the quality, responsiveness, and value you are bringing to clients. And a lessons-learned report at the conclusion of every project can identify areas for future improvement that can be built into the strategy so that future teams won't repeat the same mistakes. I have used each of these intelligence reports in the organizations that I have managed with great success and encourage you to do the same.

It is important to remember that these reports must be generated by the *front-line staff who are conducting the work* and cannot be successfully delegated to a task force or support group. This is a difficult task selling the front-line technical team on the importance of completing paperwork unless you demonstrate its value to the organization and build it into their performance. The approach used by the military is "A project is not finished until the paperwork is completed." Once this habit is formed and useful reports generated from the information, I have found that most staff will embrace these information management tools.

9.3

Intelligence reports must be generated by the front-line staff who are conducting the work and cannot be successfully delegated to a task force or support group.

Organizational Processes

The responsiveness and costs of your products and services are highly dependent on the organizational processes that support them (i.e., customer relationship management, project management, financial management). In my experience, this is the weakest link in the delivery chain of S&T-based organizations and negatively affects the cost and responsiveness in delivering products/services to clients. Improving the processes by which you deliver products/services can have a dramatic effect on competitiveness and is worth the effort. Therefore, it is important to thoroughly evaluate each of your critical processes to be sure that they contribute rather than subtract from your products value proposition. Key questions include

Is the process well defined and documented and communicated widely?

Does the process have an owner whose performance depends on its execution?

Are there competent people managing this process and know their responsibilities?

Are there control systems and key performance indicators measured?

Are staff roles and responsibilities clearly documented?

Are you getting positive customer feedback on the effectiveness and efficiency of the process?

Is the process reviewed periodically for continuous improvement?

While not as exciting as developing a novel product/service, process improvements can have a dramatic effect on the success of your strategy, since it is an area that is not traditionally emphasized by science-based organizations (i.e., your competitors).

People

Clearly, the most important evaluation is of the people in your organization that make a difference, i.e., your leadership. It is your leaders that develop and deliver your products to clients through your organizational processes. In my mind, there are three critical questions that must have satisfactory answers during the self-evaluation step.

Do we have outstanding talent leading each of our key products/services and are they happy and productive in their assignments?

Do we have a leadership pipeline of "ready now" backups for each of our key products/services?

Are our key leaders given sufficient training and coaching time to continuously develop their skills?

We all know that science-based businesses depend on innovation and our assets walk out the door each evening. Attracting and retaining top talent is the overriding secret to a successful enterprise. An S&T-based organization left to its own will tend toward mediocrity as there is always some attrition of outstanding staff while mediocre staff rarely leave. It is imperative, therefore, to be constantly on a talent search.

In evaluating your talent, it is useful to rate your key leaders using the attributes listed in Chapter 4. Identify your "A" players in leadership, management, and coaching and make sure that they have been placed in appropriate leadership roles where they can utilize their unique strengths. Identify your "B" players and prepare a realistic development plan to improve their skill sets. The remaining staff need to find other work. During the evaluation process, it is critical to wean the organization of nonperformers and ensure that the top talent is aligned with the organization's strategy and values.

☐ STEP 2. Client Feedback—Intelligence from Existing Clients on Their Current Problems and Future Opportunities and the Value You Bring to Their Organization

Once again, if you work in research and development (R&D), you always have a customer for your product or client for your services. For commercial companies, internal customers can include your S&T manager, your company's marketing department, or senior management. External clients can include companies you are selling to, venture capitalists, or angel investors. For contract research organizations (CROs) and nongovernment organizations (NGOs), clients include private and public-sector funding organizations. The concepts discussed in this chapter are quite universal in that R&D success is highly dependent on satisfying client current expectations and anticipating their future needs better than your competitors.

The best and most strategic information to help with your strategic planning efforts will come from customer or client feedback. From your current client base, you can get a pretty comprehensive picture of how well you stack up in their eyes versus your competition. It is also one of the important paths to assess market trends and stakeholder requirements. Being close to the customer was one of the key success factors in Peter's book, *In Search of Excellence, Lessons from America's Best Run Companies* [2].

Before discussing the type of feedback you need from clients, it is important to select the client base from which to survey. Most science-based organizations err on selecting too many or too few clients from which to obtain feedback. Many organizations conduct a client feedback survey on all of their completed projects, with the hope that the average result will yield a composite picture of their client's satisfaction. In other cases, there is no formal mechanism to obtain client feedback and very little interaction with clients above the level of project manager.

The first step in obtaining valuable client feedback is to select a subset of your client base that is most important to your current and future business success. I will refer to this set as *key clients*. Criteria for selection of key clients is company specific, but include measures such as volume of current business, technology leadership, market and industry leadership, and anticipated new

9.4

The first step in obtaining valuable client feedback is to select a subset of key clients from your client base that are most important to your current and future business success.

business. It is important that the list be manageable, 10–20 clients for moderate size organizations, because it takes a great deal of time and effort to build trusting long-term relationships with them. Starting with new clients, your relationship should grow from being one of many venders to a preferred provider and ultimately to a strategic partner.

Once you have selected your list of key clients, you need a codified process for building a relationship with those clients. This usually involves selecting a key client relationship manager whose role is to serve as an ambassador between the client and your company. As ambassador, he would be the face of your company, providing information and advice when asked and directing inquiries to the appropriate technical staff. He would also be the in-house expert on the client's strategy, organization, problems, and opportunities.

As a managing director, I made it a point of meeting with my key clients at least once a year along with my key client relationship manager. I would review the list of projects conducted for that client over the past year and thank the client for their business. I would then ask for feedback on three levels.

Level 1—How would you rate the quality, responsiveness, and value of the products/ services that we have provided over the past year on a scale of 1–10? How does this compare to our competition?

The scores on quality will assess your product; scores on responsiveness will assess your processes; and scores on value will evaluate the justification for your costs. Level 1 measures how well you are currently serving your clients. A key follow-up question that I have found to be extremely useful is to ask what your organization would specifically have to do to improve your scores. This is a method of extracting valuable information even when the client has given you a high rating.

Level 2—Where do you see your market going in the next few years and the products/services you will be needing from us?

This is an open-ended question that probes clients about the future and creates a dialogue where you can introduce new ideas. This can be very helpful in generating your research agenda in the strategic plan.

Level 3—What level of impact is our research having on your organization's strategy? What more can we be doing to become a strategic partner?

This question can stimulate a discussion around the more difficult problems that the client is facing or opportunities that he wishes to pursue and whether he trusts you well enough to share these with you.

The information gleaned from these client reports should be incorporated into your knowledge management system and form the basis for making strategic decisions in Step 7 of the planning process.

☐ STEP 3. Competitor Analysis—The Strengths, Weaknesses, and Strategy of Competitors' Products and Services Fighting for the Same Market Space

Make no mistake, regardless of which type of S&T-based organization you belong to, you always have competition. Commercial in-house R&D laboratories must contend with competing products or the possibility of management outsourcing their product development or R&D services to contract R&D laboratories (CROs). CROs compete with each other for both government and private sector R&D funds. NGOs compete with other NGOs for government contracts. Government R&D labs compete with each other for a piece of the federal R&D pie.

Developing a winning strategy requires that your future products/services be better than that of your competitors competing in the same marketing space. Too often, S&T managers become so enamored with the elegance of a specific strategy that they ignore benchmarking the strategy with that of their competitors. It is important to keep in mind that even an excellent strategy can be beaten by an outstanding one and a pretty good strategy will always beat a mediocre one.

Competitor analysis involves the same three steps as self-evaluation: products/ services, processes, and people. In performing the analysis and ultimately developing a strategy (see Step 7), you will want to identify areas where your strengths match up well with competitor's weakness. These represent winning themes to emphasize in both developing and selling future products or services. As part of your strategy, you will need to counteract those areas where competitor's strengths expose some of your weaknesses.

Competitor's Product/Service

Before dismissing any competitor's product/service, remember that there is a good reason why customers or clients are buying it. There are one or more features that make it better than your product/service in their eyes. In the case of a specific product, it can be quality, cost, ease of use, perceived effectiveness, or brand name. For research services, the differentiators can be even more subtle, such as a well-written proposal, established relationships, reputation of a specific principle investigator, responsiveness to a client's needs, or a proven track record of results.

The key question that needs to be answered is "what are the specific reasons why certain customers (in the case of products) or clients (in the case of services) purchase a competitor's product/service over ours?" Answers can be found in examining the product or service itself (through reverse engineering) and obtaining feedback from customers that use their products or services.

While it is useful to compare your current product's success with that of your competitor's, it is even more important from a strategic point of view to predict the relative success of your future products and services in the pipeline to that of your competitor's. This is a more difficult exercise as the information is harder to get (but not impossible). Much information on new products can be gleaned from the literature, patent searches, and your competitor's own product literature.

Competitor's Organizational Processes

The best way to evaluate a competitor's processes in relation to yours is to measure relative outcomes. Do they produce their product faster, cheaper, and of a higher quality than you do? This is a good indication of the competitiveness of their manufacturing process. Do they provide a more responsive and value-added research service than you do? This is a good indication of their project management process. Do they win more research projects than you do? This is an indication of the quality of their proposals. Do clients prefer doing business with them? This is a good indication of their client relationship management process.

As mentioned previously, a good knowledge management system that you can refer to during the planning process is invaluable as the information needed is normally generated over a period of years by a broad spectrum of staff. When conducting a customer or client review, questions concerning your organization's quality, responsiveness, price, and value are asked in a relevant context; i.e., compared with each of your competitors. The data and information obtained in these reports become useful for both self-assessment and competitor analysis.

Competitor's People

As discussed previously, the key differentiator in any science-based organization is the quality and motivation of its leadership. Leaders develop the products, oversee the services, and manage the processes by which the organization competes in the marketplace. I have always felt that over the long run, science-based organizations compete not on the basis of their products or services but on the quality of their leadership.

I know that in the R&D services business, sophisticated clients buy people not organizations. So regardless of the money spent on company branding and advertising, it's the specific people in the organization that attract business. When managing a contract biopharmaceutical development laboratory, my clients would ask for specific study directors by name as a condition of getting their business.

It is a good practice to get to know your competitor's key leadership as well as you know your own. I routinely made it a practice to take one or more of my competitor's key leadership to lunch at least once a year (often at a technical conference). We would often share stories about common clients and issues that we were facing in our own companies. While protecting company secrets, we were willing to give each other useful information that made these meeting worthwhile.

My motive was always based on recruiting. I would ask my competitor how happy he was in his current position and whether or not he would be interested in a job offer from our company. The common response was that they were very happy with their current jobs. Those were the very people I was interested in recruiting: talented, successful leaders who were happy with their work. Inevitably over the years, situations changed, and I was able to hire several of these recruits along with the clients that they brought with them.

☐ STEP 4. Technology Trends—An Awareness of Competing and Emerging Technologies That Could Disrupt the Strategy

As discussed previously, R&D product life cycles are becoming increasingly shorter, and today's leading products can become obsolete in a short period of time. It is therefore critically important to develop an awareness of technology trends that could affect your current and future S&T strategy. For purpose of discussion, I have divided these trends into three parts: incremental, evolutionary, and revolutionary or disruptive.

The first and easiest trend to monitor is the incremental one; the relative rate at which both you and your competitors are improving your existing products and services (i.e., making them better, faster, and cheaper). Most R&D groups spend a considerable amount of resources (both staff time and money) in protecting their current product lines from competitor's advances (as well as they should) by product differentiation (improving their product's features) and process improvements (to produce their product faster and cheaper than their competition). When done well, this can extend the life cycle of a product by several years.

As an example, I was asked to turn around an environmental research laboratory that had experienced such a trend. It found itself losing several proposals for large environmental monitoring and assessment projects despite having superior technical reports recognized by their clients. In debriefings with clients, I found that competitor's bids were substantially lower in cost. After overcoming the initial reaction of denial, we examined several processes and found that nearly half of our costs were associated with the management and quality control of large data sets. This was at the time when the trend in automated data collection and information management systems was emerging. We decided to invest internal resources at considerable expense on a state-of-the-art environmental information management system and were able to reduce our costs by a factor of 3 while improving overall data quality. These improvements resulted in recapturing most of that business and extending that product line for several more years.

When evaluating incremental technology trends, it is important to benchmark not only the relative competitiveness of existing products but also the relative rate of improvement in product quality, cost, and speed to market. If you find that your current products are behind your competitors and your rate of product/process improvement slower, it may be time to either abandon that product or rely on your research team for a transformational technology to leapfrog your competition.

Depending upon your industry, a sizable portion of your research portfolio may be allocated to projects that target an evolutionary change in your product or service. Unlike incremental change, an evolutionary change involves the development of a new technology or marketing business model that results in an immediate and substantial competitive advantage. As mentioned in Chapter 5, justifying R&D spending that targets evolutionary change is challenging as most of these projects involve considerable risk. Financial managers are much more comfortable justifying return on investment calculations on incremental improvements. This is where having a balanced R&D portfolio consisting of projects with both short-term (incremental) and long-term (evolutionary) results make sense.

An example of an evolutionary product that we were fortunate to have developed at Arthur D. Little was the first practical phototoxicity and carcinogenicity evaluation using novel instrumentation for our client, Johnson and Johnson. We developed this instrument and used it to evaluate the toxicological effects of Retin-A and gained approval for its use by the Food and Drug Administration (FDA). This breakthrough sets the standard for phototoxicity testing in the United States and helped us corner the market for several years.

Another example involved developing a combination of analytical tools (mass spectrometry and biomarker analysis) and reference materials to definitively characterize oil constituents (pure and weathered) in the environment, which our scientists named "environmental forensics." This expertise was used to track oil spills and assign potential culpability and was used extensively in the Exxon Valdes oil spill. This analysis transformed the process of Natural Resource Damage Assessment for oil spills internationally.

The final trend that needs to be monitored is the revolutionary or disruptive technology trend. Seemingly irrelevant emerging technologies can be innovatively applied to render your product, service, and even your company obsolete. A good example of this is the Polaroid Corporation, which for decades had a monopoly on instant photography based on hundreds of iron-clad patents on photochemistry. With the introduction of digital photography, instant photography became a commodity taking Polaroid from a market leader to essentially out of business. Similarly, Kodak, who had a commanding market share of the photographic film market, found its film business disappear in a matter of a few years with the advent of digital photography.

There are currently several emerging technologies that appear to be on an exponential growth curve that will have a dramatic effect on practically every market segment. These include nanotechnology, biotechnology, bioinformatics, bioengineering, sensor technology, robotics, 3D printing, artificial intelligence, advanced computational systems, and remote-control technology.

For an excellent and uplifting read on how many of these technologies may make this a better world, I recommend reading *Abundance* by Peter H. Diamandis and Steven Kotler [3]. Each of these technologies has the potential of disrupting many existing technology-based companies and need to be constantly monitored.

The overall process of digitalization is fundamentally changing the way that business is conducted globally. When reviewing your processes and that of your competitors, it is critical to stay ahead of the curve as digitalization will make or break your strategy.

Typically, such information on these disruptive trends is obtained from a variety of sources, including technical conferences, patent searches, journal papers, and the like. In addition to monitoring your own technology, you need to look beyond your field of specialization to develop both the depth and breadth of intelligence which leads to innovation. As mentioned in the beginning of this chapter, this is an ongoing process conducted by a variety of organization's leadership, including executives, staff scientists, program managers, and marketing managers. Developing a knowledge management system that captures this information institutionally will help immensely when it comes to Step 7, strategy synthesis.

9.5

There are currently several emerging technologies that appear to be on an exponential growth curve that will have a dramatic effect on practically every market segment.

☐ STEP 5. Market Analysis—A Thorough Knowledge of Market and Industry Trends

In our discussion of competitor analysis, the focus was on obtaining detailed actionable intelligence on our direct competitors. The focus of a market analysis is to broaden that discussion on the industry to determine market trends that affect our decision making.

There are two types of market data that can provide useful information that fills in the blanks in your competitor analysis; published data and field data, sometimes referred to as secondary and primary data. Since published or secondary data is easier to obtain, there is a tendency to rely on it heavily. Primary data is gathered from interviews with industry participants and observers. It is much more time consuming and oftentimes results in conflicting data but generates the most useful information when formulating your strategy. Once again, emphasis should be placed on obtaining this industry data continuously as part of your overall knowledge management process and updating it regularly.

There are numerous sources of published data. Starting from scratch, one should determine the firms in a specific industry, especially the leading players starting with the industry's *Standard Industrial Classification code,* from the Census Bureau's *Standard Industrial Classification Manual.* Comprehensive studies of specific industries are also available that can give a broad overview as well as a good source for additional research. There tend to be comprehensive studies conducted by economists or more focused studies by securities or consulting firms. Trade associations, trade magazines, and the business press can oftentimes be a good source of information if they are read over long periods of time so that trends can be teased out. Finally, company documents such as annual reports, Security and Exchange Commission form 10-Ks, proxy statements, and prospectuses should not be overlooked. Also of value are executive speeches, press releases, product literature, and patent filings.

As mentioned, generating field data or primary data is more difficult and time consuming as it involves interviewing important industry sources. Industry observers are numerous and include standard setting organizations (e.g., underwriter's laboratory), unions, the press, local chamber of commerce, state and federal government, watchdog groups (e.g., Consumers Union), the financial community, and regulatory agencies. Service organizations are also a good source of information and include trade associations, investment banks, consultants, auditors, commercial banks, and advertising agencies.

Given a broad list of potential sources of information, it is important to have a strategy. Initially, try and get an unbiased view of the industry by interested third parties; those knowledgeable about the industry but do not have a competitive or direct economic stake in it. They are also a good source of direct industry participants to whom it is worth talking. Another good starting point is individuals who have been quoted in articles and speakers at industry conventions. Oftentimes, it is worth the effort to get introductions to such speakers form a referral. For more detailed information on market analysis, I recommend Porter's excellent book, *Competitor Analysis* [4].

Generating an industry analysis seems like a daunting task. It requires significant resources and time and fraught with conflicting information. Therefore, it is important that all the organization's leadership be involved in generating market data. At quarterly strategic management meetings, which we will discuss in the next chapter, time should always be set aside to discuss and update market trends with information obtained from sales and marketing, R&D, product managers, and the like.

At this point, you may be wondering why an S&T manager needs to focus attention on market trends. Market and technology trends are inextricably linked. Old models relying on just technology push or market pull are becoming outdated. In developing an S&T strategy, one needs to consider not only the effect that technology will have on the marketplace but also the way in which market trends will affect the acceptance of technology. The two examples from the Biotechnology and Oil and Gas industries illustrate this point.

9.6

Market and technology trends are inextricably linked.

I asked a leading medical device expert for an overview of the biotechnology market. In the pharmaceutical industry, biotechnology drugs have revolutionized the treatment of many diseases, including cancer, cardiovascular disease, and autoimmune disease. These drugs are often delivered parentally, either through an intravenous or through a subcutaneous injection, to avoid degradation in the digestive tract. To help ease the burden of injection, the delivery of these drugs has evolved from a syringe filled from a vial, to a prefilled syringe, to an autoinjector, which automatically depresses the plunger to deliver the drugs.

As the biotech industry continues to develop novel therapies to treat more and more patients, the patient experience is becoming an area of increased focus. In researching the patient experience, several challenges in the existing environment have been identified. For instance, many patients are classified as "needle-phobic" and will not accept a therapy requiring them to inject themselves. Others will rely on their doctors' offices to administer the injection, taking up valuable resources from the healthcare system.

Additionally, as biotech discovery and development become more sophisticated, new therapeutic targets are being identified that require much larger doses of drugs. To deliver much larger doses, a higher volume of the drug must be delivered, or the drug must be highly concentrated, greatly increasing the viscosity. Both options create a much more difficult injection, increasing the duration and the force required. Existing injection systems will be inadequate for these "next-generation" biotech drugs.

Because of these dynamics, the pharmaceutical industry is investigating the market for novel delivery technologies, including implantable and wearable devices. Making an injection easier is expected to significantly improve the patient experience. In the long run, these improvements may allow for some drugs that currently require an IV to be delivered by an injection by a patient at home, reducing hospitalization durations. In addition, a great deal of research is being focused on oral biologics, which will allow for the delivery of these complex molecules by a pill.

I also asked one of my colleagues for some feedback on the recent IHS CERAWEEK conference (Cambridge Energy Research Associates) in Houston on *Energy Transition: Strategies for a New World*. He was not surprised to find that peak oil was not on the agenda at the conference. Front and center in the agenda, as you might suspect, was discussion on when will oil prices recover and what can be done to expedite the recovery. The mood was dour, and for most, there was resignation to a lower longer outlook. In recent times, the price of oil has recovered and that combined with the International Energy Agency, proclaiming that oil prices appear to have bottomed out, has at least pacified those who feared $10-barrel oil.

Indeed, while the recent rise in price is good news for many quarters, it would be a serious mistake to believe that we have survived another cycle and that we are on an inevitable climb to $100-barrel oil. As with the other cycles, this one has unique aspects that will impact energy planning for the future. Just as $25-barrel oil was undervalued, so too was a $100-barrel oil unsustainable. Oil will not be replaced by renewables in the foreseeable future, but a more balanced energy portfolio composed of solar, wind, nuclear, and oil (conventional and nonconventional) will compete against each other for market share.

Organization of Petroleum Exporting Countries (OPEC) doesn't have the influence it once had in the preshale world. Production is dropping in the United States, and it is estimated that another 600,000 barrels a day will go offline this year. Many in OPEC never wanted sustained prices above $100 a barrel, because they feared exactly what has happened. The cost of production has dropped because of innovation. The industry is restructuring and will hesitate to begin drilling again until they see some stability in price and at least for $50-barrel oil. It is interesting to note that the low price of oil may see the majors acquiring a greater presence in shale production as they purchase overextended independents.

Many suspect that the recent price rise is due to expectations of a political agreement being pursued by Russia and Kingdom of Saudi Arabia, but no real recovery will occur until demand eats away at the oversupply. A recovery of the China economy would go a long way to accelerate the demand, but the more serious prognosticators do not expect a sustained price recovery for another 6–18 months. Even then, the dynamics of the market have changed most notably again in the shale production. While shale production is expected to drop again this year, it can be turned back on relatively quickly, which would temper any sustained resurgence. The increased competition also will temper demand, and more volatility in the market may become the norm. It was a simpler time when the peak oil advocates thought they had it all figured out.

As it turns out, we have seen oil prices fluctuate continuously and are now hovering at the $50–$60 dollar a barrel range. Given the current global politics including the rejection of the Iran agreement on nuclear weapons and the pending agreement with North Korea, it is easy to see how resource based technology organizations need to keep a close watch on emerging trends.

There are many more examples that can be explored.

In the health industry—Disease prevention through a healthy lifestyle; nutrition and genomics; disease cures through stem cells; personalized medicine

In the energy industry—General acceptance by governments and industry on global warming; increase in petroleum reserves due to fracking; viability of renewable energy sources such as wind and solar

In the transportation industry—Automobiles transitioning from fast horse and buggies to self-driving mobile computers; robotics changing the face of deliveries

In the food industry—The whole food movement; Genetically modified organisms (GMO)'s beauty or beast; the role of pesticides in global food productivity

In the manufacturing industry—Robotics and 3D printing

In the construction and building industry—Smart buildings; advanced materials

In the water and wastewater industry—Desalinization; wastewater reuse

The two important takeaway lessons when developing a strategy is that you must first look at both the microlevel through competitor analysis and the macrolevel through market and industry analysis. And second, you must factor in the interdependence between technology trends and market trends.

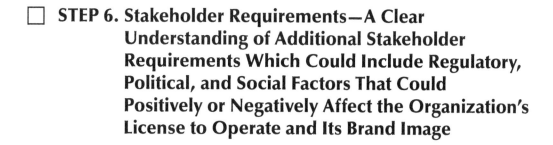

STEP 6. Stakeholder Requirements—A Clear Understanding of Additional Stakeholder Requirements Which Could Include Regulatory, Political, and Social Factors That Could Positively or Negatively Affect the Organization's License to Operate and Its Brand Image

We have elaborated on the first five steps of the data collection process: self-assessment; competitor analysis, client feedback, technology trends; and market analysis. We will now discuss the final step before synthesizing the data; the importance of factoring in stakeholder requirements into the strategic planning process with the focus on three critical stakeholders: regulators, employees, and the community in which your organization impacts.

In the case of regulators, the sense of importance and urgency is acute. Meeting regulatory requirements such as those required by the Environmental Protection Agency or the FDA can mean the difference between success or failure of a new product or in the case of a startup company, the business itself. A successful strategy must not only consider the competitive advantage of a new product but also its ability to pass muster with the regulatory agency. In the case of the FDA, a new product must meet a threshold efficacy requirement while also providing a margin of safety. In my experience working with dozens of biotechnology clients, I have seen two widely different philosophies and approaches to developing a regulatory strategy.

Some clients view their relationship with the FDA as an adversarial one and the regulatory requirements and process as a barrier to overcome. Regulatory experts are hired to "guide" the development process that presents the minimum amount of data necessary to obtain approval in as fast a time as possible. The attitude is that no matter how much information we provide to the agency; they will always be looking for more. On the other hand, I have worked with clients who view the FDA as collaborators and spend a great deal of time upfront in trying to understand what the issues of concern are and how to generate the data to address them.

In my experience, these two strategies mimic the tortoise and hare story. In the adversarial relationship, product data does get reviewed faster but often questions are raised that require additional research and development. These iterations wind up taking much more approval time and costs than if these questions were identified early on and addressed. Taking the time to codesign a development strategy with the FDA study group results in a more comprehensive study plan with a greater chance of initial success. While it may seem to cost more and take more time initially, in the long run it will likely be the most cost effective and fastest route to approval. An important side effect of such a strategy is the development of trust among the regulators about your company's commitment to safe and efficacious products. Such an approach will require balancing impatient investor desires with the need for a thorough development strategy.

9.7

Taking the time to codesign a development strategy with the FDA study group results in a more comprehensive study plan with a greater chance of initial success.

As an example, we worked closely with Johnson and Johnson on their development and regulatory approval of Retin A. From the very beginning, a strategy was developed that focused on answering every question that we could come up with from experimental design to interpretation of the data. This resulted in developing a novel apparatus for exposing nude mice to a carefully metered dose of light, which became the industry standard for FDA approval.

Although less urgent, but equally important is to incorporate the needs of your employees and community in strategic planning. The concept of corporate social responsibility is relatively new, but I believe will become an increasingly important aspect of an organization's strategy and vital to its long-term viability. Today's newspapers are filled with stories about corporate greed, CEO salaries, corruption, and income inequality to the point where capitalism's excesses are being openly protested by the public. Despite capitalism's powerful economic model that provides financial incentives for solving problems and creating opportunities, the basic goal of maximizing shareholder value by all major corporations is being challenged. A more comprehensive model for business strategy based on a balanced scorecard [5] was introduced over 20 years ago, and while many organizations profess to practice it, few in my opinion have embraced it strategically.

Corporate social responsibility is the belief that business is not divorced from the rest of the society and a company should consider the social, ethical, and environmental effects of its activities to improve the quality of life of its staff and the community around it. As part of a corporate strategy, the goal becomes to balance the needs of all stakeholders: a good return on investment for shareholders, fair and equitable compensation for management and staff, and environmentally sustainable products and industrial processes. Such a strategy will inevitably result in a dramatic improvement in an organization's brand and have a bottom-line payoff.

Unfortunately, while many of today's corporations have publicly endorsed the concept of corporate social responsibility, it has been given lip service for the most part. The trend however is moving more towards sustainable products, whole foods, fair trade practices, and the like. Enlightened companies are realizing that consumers are looking to buy from socially responsible companies and that this represents a competitive edge.

I have had direct experience with the benefits of such a strategy while managing an industrial S&T organization embedded in a small residential community. We were determined to be a good corporate neighbor and incorporated into our strategy several community initiatives. Being a science-based organization, we decided that contributing to science education in the local school system would have the most impact. We organized and managed an annual science fair in collaboration with the community's high school teachers. Many of our senior scientists volunteered their time to judge the exhibits. Very little money was required as most of the effort was pro bono on the part of our staff. Over the years, the science fair became a major annual event that enhanced our image as a good corporate citizen.

In addition to the science fair, there were many other initiatives. Our chief financial officer volunteered his time on the town's finance committee. Our human resource manager organized several charitable events. We volunteered the use of our land for the town to hold summer concerts on the lawn. Over time, we built a reputation as a good corporate neighbor. When we petitioned the town for an expansion of our facilities, it passed with flying colors as we were a trusted member of the community.

☐ STEP 7. Strategy Synthesis—Collecting and Synthesizing Disparate Data and Information into Actionable Intelligence on Which to Make Decisions

Armed with the information obtained in the first six steps of the planning process, an organization's leadership team can then develop a business strategy consisting of several strategic thrusts designed to allocate its limited resources to protect and exploit its technologies; compete in a favorable market space; and recruit and retain top-level talent. The most important focus of the strategic planning process is making decisions on where to allocate scarce resources. Too many organizations get bogged down on the process and writing the plan and lose sight of the real importance of the planning process, which is to make tough decisions about creating an exciting future and where to allocate company resources. This is the time to be thinking out of the box while challenging every assumption. Before starting any strategic planning process, I highly recommend that you and your team read and discuss *Blue Ocean Strategy* [6] and *Big Think Strategy* [7].

Mission, Vision, and Values

Before discussing strategy, it is important to clearly define the organization's mission (why do we exist?), vision (what do we want to become?) as well as values (how should we behave?) that will ultimately determine its culture.

The mission is usually a variation of developing, deploying, and exploiting S&T to create value for the organization depending on the industry sector. In the case of government research institutions, the mission is often expanded to include recommending policies and commercializing technologies based on sound science to stimulate economic and social development.

The vision describes the aspirations of the organization that defines success. It is usually defined in business terms such as capturing market share, growth in revenues, and creating shareholder value. Equally important is how the organization will improve the human condition through the development and exploitation of S&T.

Organizational values are then developed that identify the behaviors necessary to support the mission and vision. This often-overlooked step is critical to shaping the organizational culture essential for success. Johnson and Johnson's Credo shown later exemplifies the balanced scorecard approach and is one of the best value statements I have ever read that captures its mission and vision. Having had Johnson and Johnson as a client, I can attest to the importance of the values in the Credo. Whenever there was a difference of opinion or point of contention as to a specific decision to make or direction to pursue, The Credo would be pulled out and read aloud to help the competing staff make the best decision possible.

Developing a Strategic Agenda

One you have finalized your mission, vision, and values, a strategy can be developed to fulfill your mission and achieve your vision. As stated previously, the essence of a winning strategy is to make decisions about what to focus on and where to invest your limited resources. This will result in a competitive advantage, given the information and data gathered in the first six steps of the planning process. I prefer the term strategic agenda rather than strategy, as there are several strategic thrusts that must be simultaneously pursued to achieve one's vision (Figure 9.2).

On the basis of the information obtained in Steps 1–3, you should have a pretty good idea about your current state; i.e., where your products/services stand with your client base versus your competition. Your vision statement provides you with a clear idea of a proposed future, desired state. A strategic agenda represents a road map that takes you from your current state to the desired state. You must identify specific thrusts to pursue that closes the gap between your current and future states. The thrust areas should address your current and future clients, products/services, processes, staff, and stakeholders and have specific identifiable outcomes that can be measured. Before we elaborate on the thrust areas themselves, there are several general comments that need to be taken into consideration.

First and foremost is to determine the degree of change that may be necessary to get from the current state to the desired state. If you are currently an industry leader, you may only be looking at an incremental change to effectively maintain your position in the marketplace. If you sense that there are pending technology trends, market trends, or competitor initiatives that may threaten your leadership position, you may need to plan for a more progressive, evolutionary change. If you are not in a position of leadership and see your business threatened by any of the points mentioned earlier, you may need to develop a revolutionary approach to transform your current business in dramatic ways. Clearly, when moving from incremental to evolutionary to revolutionary change, the thrust areas become more difficult and the risk level increases dramatically. Having led several "turnarounds" requiring transformational change, I can attest to their difficulty.

Next, it is important to place your strategic thrusts and resulting outcomes in context. Your competition also has their strategy with which to beat you in the marketplace, so you must measure your progress versus your competitor's, both relative to your current competitive position and the relative rate at which you are both improving.

Another consideration is to consider the ability of the organization to successfully implement the strategic thrusts by critically examining the self-assessment developed in Step 1. Many good strategies have failed due to a lack of appreciation for the difficulty in implementation, particularly during a transformational change. The barriers are many, including organizational policies and culture, leadership talent and commitment, competitor countermoves, and lack of anticipated resources.

One way to mitigate surprises during the implementation of your strategic agenda is to develop several scenarios that test the robustness of the strategy. I have found these "what-if" scenarios to be useful in terms of developing work around contingency plans that can be incorporated into the planning effort. At least three scenarios should be developed: an optimistic scenario that represents the successful implementation of each of your strategic thrusts: a pessimistic scenario driven by disruptive new technologies, market trends, regulatory hurdles, or organizational barriers, and a most likely scenario in the middle. An additional benefit from scenario development and contingency planning is the consensus that it can build among the leadership team involved in the planning effort.

9.8

One way to mitigate surprises during the implementation of your strategic agenda is to develop several scenarios that test the robustness of the strategy.

In developing a strategic agenda, the focus will be on five imperatives: a market strategy, regulatory strategy, key stakeholder strategy (outside-in approach), a product strategy, and a resource strategy (inside-out approach). For each of these strategies, specific strategic thrusts need to be identified that, if successfully executed, will lead to achieving your vision and mission. In choosing each of your strategic thrusts, be sure to focus your investments (people and financial resources) on what you consider the few critical success factors. Too often, strategies are loaded with initiatives that dilute funding and talent and distract from critical initiatives. Strategy should not only describe what you plan on doing but also on what you will not be doing.

JOHNSON & JOHNSON CREDO

We believe our first responsibility is to the doctors, nurses, and patients, to mothers and fathers and all others who use our products and services
In meeting their needs everything we do must be of high quality.
We must constantly strive to reduce our costs in order to maintain reasonable prices.
Customer's orders must be serviced promptly and accurately.
Our suppliers and distributors must have an opportunity to make a fair profit.

We are responsible to our employees, the men and woman who work for us throughout the world.
Everyone must be considered as an individual.
We must respect their dignity and recognize their merit.
They must have a sense of security in their jobs.
Compensation must be fair and adequate, and working conditions clean, orderly, and safe.
We must be mindful of ways to help our employees fulfill their family responsibilities.
Employees must feel free to make suggestions and complaints.
There must be equal opportunity for employment, development, and advancement for those qualified.
We must provide competent management and their actions must be just and ethical.

We are responsible to the communities in which we live and work and to the world community as well.
We must be good citizens—support good works and charities and bear our fair share of taxes.
We must encourage civic improvements and better health and education
We must maintain in good order the property that we are privileged to use, protecting the environment and natural resources.

Our final responsibility is to our stockholders.
Business must make a sound profit.
We must experiment with new ideas.
Research must be carried on, innovative programs developed, and mistakes paid for.
New equipment must be purchased, new facilities provided, and new products launched
Reserves must be created to provide for adverse times
When we operate from these principles, the stockholders should realize a fair return.

Market Strategy

The success of your market strategy is directly dependent on the quality of your competitive intelligence. In Steps 1–6 of the strategic planning process, data was gathered on self-assessment, competitor analysis, client feedback, technology trends, market analysis, and stakeholder requirements. The data in and of itself is relatively useless unless it is used to develop competitive intelligence. Gathering data is only the first step in the process and represents the raw material from which to develop information. Information is the grouping of data to form comparisons that reveal a larger picture and provides more meaning. Competitive intelligence is the goal of the planning process that reveals critical information that provides insights for competitive advantage and forms the basis of strategic decisions and actions.

For a simple example, the annual revenues of a competitor represent a data point. Revenue growth rate over the past 3 years represents information. Nonorganic revenue growth rate caused by several mergers and acquisitions represents competitive intelligence that may indicate a vulnerability that could be exploited. The goal in developing a market strategy is to turn data into information and ultimately competitive intelligence that opens patterns or trends on which to make decisions on a course of action.

9.9

Competitive intelligence reveals critical information that provides insights for competitive advantage and forms the basis of strategic decisions and actions.

A winning market strategy is the one that allows you to successfully compete in a favorable, highly attractive market segment with a subset of clients who highly value your products/services. In identifying a highly attractive market segment, there are many factors to consider, including the size and growth rate of the sector; profitability of its client base; its diversity and sustainability; sensitivity to price and service; the number and strength of competitors; barriers to entry; maturity and complexity of the segment; and economies of scale. In the case of the biotechnology industry, a market segment could be a disease state being targeted for cure.

A second consideration is your ability to successfully compete within that market segment. Factors to consider include the technical capabilities of your staff; regulatory knowledge; patent protection; alignment of your Internal Research and Development (IR&D); applicable facilities and equipment; subcontractor and supplier access; reputation and track record within the market segment; and potential financial returns on investments. The figure given later represents a matrix tool from which to make decisions (for more detail, see the Competitive Forces Model by Porter and the Business Portfolio Analysis by BCG, McKinsey and GE, http://www.valuebasedmanagement.net/methods_ge_mckinsey.html). As stated previously, the key to making decisions about which market segment to compete in should be based on the competitive intelligence you have gathered during the planning process (Figure 9.3).

As a senior manager in contract R&D and technology consulting organizations, I have had the good fortune to have worked in several different market segments, including the US Federal government, several state governments, the pharmaceutical and biotech industry, the oil and gas industry, and the chemical and agrochemical industry. Each of these markets represented unique challenges in developing a winning strategy, and the market segmentation matrix is a good place to start.

Once a market segment has been chosen, the next step is to choose the most attractive clients within that market segment. Client selection is one of the more important decision that gets made during the strategic planning process and is often determined by the feedback from current and potential client accounts. Account management is a discipline that focuses on the selection of high-value "key" accounts within a market sector where you can successfully compete. Key accounts are usually chosen based on their potential for growth and profitability of your current and future products/services.

Market Segmentation Matrix

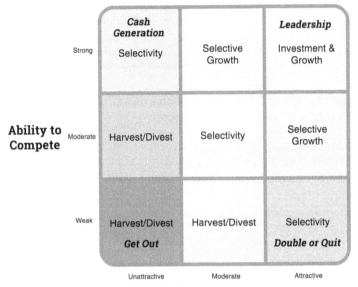

FIGURE 9.3 Market segmentation matrix.

One of the most common strategic thrusts identified in a market strategy is to build relationships with each key account so that you can transition from being just a vendor to a preferred provider and ultimately into a strategic partner. In selecting a key account, it is important to determine first what their relationship is with one or more of your competitors. When this relationship is strong, this may not be a good key account choice, regardless of its attractiveness. Conducting R&D is an inherently risky business where clients invest in your ideas without guaranteed results. It is best to identify not only the account's attractiveness but also the value you may bring to the relationship above and beyond your competitors. At the very least, you want to be a preferred provider so that, when a need arises, you are one of a very few organizations that are invited to present a proposal.

Your goal is to become a strategic S&T partner. This is easier said than done and requires a substantial commitment of time and resources. It usually requires a combination of a successful track record on previous projects; a dedicated relationship manager that maintains contact in between active projects; and the ability to provide sound advice on the client's marketplace and the R&D support needed. In my experience, most S&T organizations do not view key account management as strategic and give it just lip service. However, it is one of the most important strategic decisions you can make. Using an example of the Pareto principle, 80% of most client revenue will come from 20% of the client base. By focusing your marketing investments (and in some cases your IR&D investments) on key clients, you will most likely increase your return on investment dramatically.

As an example, our organization once had the opportunity to bid on an R&D development project for a key client, knowing that there would be a large follow-on project for implementation. By adding some of our own IR&D funds to the bid, we could successfully win the project. By "super pleasing" the client, we could procure the much larger follow-on project. While this decision seems obvious after the fact, the competition for IR&D funds was fierce and the decision to allocate them to this project was based on the priority of this key client. By strategically identifying your key clients and emphasizing their importance to your organization, you can focus more attention not only of your R&D staff but also of your entire organization, including the responsiveness of key managers and administrative staff.

Once the decisions are made on the target market segments and key accounts, a market strategy can then be developed that allocates critical resources and management attention to those accounts. Generic examples of strategic thrusts within a market strategy could be the following:

- Evaluate the potential of entering market segment X by opening a local office

- Triple the revenues on key account Y by developing new products/services that help them increase market share.

- Obtain a seat at the annual R&D strategy meeting of client Z

Each of these thrusts will require action to be taken and accountability to be measured.

Product Strategy

As an R&D director or manager, your primary responsibility will be to make strategic decisions that create value from the science that you manage. This is particularly difficult in S&T-driven organizations. The business results of your decisions oftentimes are years into the future and loaded with technical and market uncertainty. In today's R&D ecosystem, in addition to your in-house R&D, decisions need to be made on partnerships, alliances, or acquisitions to achieve product objectives. Innovation becomes a business process as well as a research process.

The first step in this process is to develop an overarching technology strategy. This is a vital component of technology-based businesses and requires the translation of the organization's business strategy to determine the precise role that R&D will play in its success. How will you create new value from R&D through the development and renewal of technical products and services? What emphasis will you place on new products? Can you develop a technology platform that contains intellectual property and know how that will allow multiple product development streams? Will you develop new technologies in-house or obtain them from outside your organization? What are the competencies that are critical among your technical staff and do you have sufficient breadth and depth? What alliances should you be pursuing? Significant time should be spent by your R&D leadership team in answering these questions before proceeding to Step 2.

Once an overall technology strategy has been developed that supports the organization's business strategy, Step 2 will be to develop a portfolio strategy that creates products and services that support the organization's market strategy and resource strategy. Executing such a strategy and delivering products and services to clients is the essence of the business and must be based on providing sound value to clients. There are only two truly foolproof product strategies: either produce a desirable product/service that no one else has or produce a competitive product at the lowest price. While highly desirable, both strategies are hard to achieve. Novel products are quickly reverse engineered and copied by aggressive competitors. Lean manufacturing processes are desirable, but high-quality research products and services are rarely delivered at a low lost. Competing on price is a last resort for S&T organizations and often leads to panic investments in new products and services to recapture profits.

The aim of most S&T strategies is to produce the highest value for customers. This is a combination of product features, benefits, costs, and service to meet client needs better than your competitors. Developing such a strategy involves the generation of intelligence from the data collected during client feedback, competitive intelligence, and technology trends. Given the shortened life cycles of most products and services these days (usually 12–18 months), an S&T strategy usually involves the development of a product portfolio consisting of several technical development projects. These projects run the gamut from low-risk short-term projects to extend the life cycle of current products and services to more complex higher risk, long-term projects to develop newer, more competitive products of higher value to clients, particularly key clients identified in your target market sector.

A well-designed portfolio strategy incorporates the technology goals of the organization and manages resources (technical staff, capital, and expensive technical equipment) as efficiently as possible. Portfolio decisions are made based on what produces the most value to the organization and resources are parsed accordingly.

Given the uncertain nature of S&T, developing and managing an R&D portfolio is critical to the success of your organizations' overall strategy. It is important therefore to develop decision criteria for the selection of the technical projects to include in your R&D portfolio that supports your organization's financial strategy.

The third step in developing an S&T strategy is to select the most promising projects in your portfolio that meet both the short- and long-term needs of the organization. The selection process involves an initial screening of competing R&D concepts from both internal and external sources. This is a universal process whether you are a client evaluating a concept from a contract R&D organization, a venture capitalist evaluating an R&D business concept, or an R&D manager evaluating your staff's concepts. While each organization will have different research agendas and financial resources, the following are questions that can be used to initially screen concepts for further evaluation.

1. Does the concept have strategic value?

Will the research proposed make a substantial contribution to the organization's product strategy?

Is the proposed research a response to a key client request? If so:

How committed is the client to solving this problem/pursuing this opportunity?

What are the benefits of the concept to the client?

Is the client an advocate of the concept/technical approach?

Is the client willing to fund a well-written proposal?

Do the ideas presented in the concept have the potential for intellectual property development and patent protection with commercialization potential?

Is the concept innovative enough to lead to product leadership in the marketplace?

2. Does the concept have technical merit?

Is the problem or opportunity well defined?

Do the objectives of the proposed project solve the stated problem or create the proposed opportunity?

Does the technical approach and experimental design support the objectives?

Is the research methodology based on sound science building on existing research knowledge or emerging technology?

Do the research staff/consultants have the appropriate technical expertise and experience?

3. Can the concept be properly executed?

Does the proposed project leader have the management experience and track record to lead the project?

Is the work plan well defined and sufficiently detailed to ensure appropriate monitoring of the work flow and deliverables?

Is the project schedule realistic?

Is the proposed level of effort of key staff sufficient to accomplish the stated tasks?

Is the proposed budget sufficient to accomplish the work?

Are the proposed key staff available for the level of effort proposed?

Will the needed facilities and equipment be available at the designated start of the project?

4. What are the potential risks?

What are the potential major risks (political, technical, environmental, safety, and health) that could affect the organization's reputation and/or liability?

What are the management actions and preventative measures planned to mitigate the risks?

The screening process can be used to quickly reduce the concepts to a manageable number from which to request more detailed proposals. A more rigorous technical, management, and financial review can then be conducted to select the projects to be included in the S&T portfolio.

It is important at this stage to carefully review the self-assessment conducted during the strategy review process. It is easy to become enamored with an exciting concept and fail to sufficiently question the organization's capacity to execute such a project. Can the organization commit to the identified resources? Are there single points of failure such as a key principle investigator with no qualified backup? Is the project team on the same page? Are there sufficient, well-defined milestones listed to evaluate the project's progress?

A good analogy to developing an S&T portfolio is the concept of a balanced financial portfolio. A balanced R&D portfolio is one that best manages risk. For short term, low-risk, projects are chosen that extend the existing product lifespan through increased quality, speed, and/or decreased cost. For long term, high-risk projects are chosen that develop new products or new markets for modified existing products. The degree of overall risk and portfolio of projects will depend on your current self-assessment, financial strategy, and the ability to achieve your mission and vision.

For an excellent overview of R&D strategy, I recommend *The Smart Organization, Creating Value through Strategic R&D* by David and Jim Matheson [1].

Resource Strategy

The three critical resources needed to execute your overall strategy are key technical staff, capital, and scientific facilities and equipment. A shortfall in any of these three resources will jeopardize your ability to successfully execute your strategy despite the elegance of your strategic plan.

A Staffing Strategy That Ensures Sufficient Depth of Talented and Motivated Staff

I know I'm stating the obvious when I say that the most important resource and critical success factor in any S&T organization is the breadth and depth of its technical staff. However, I have too often seen this critical resource either taken for granted or ignored when a strategic agenda is developed. Some of the key questions that need to be answered are as follows:

Does the organization have the key technical staff to execute the research portfolio developed in the product strategy?

Are the key staff totally aligned with the research agenda being proposed and committed to its execution?

Have the key staff been trained in the organization's execution processes?

Is there sufficient depth to cover for any attrition of key staff?

If the answer to any of these questions is no or maybe, this is a red flag and warrants a strategic thrust in the resource strategy. For example, if the execution of a key project/program depends on a new technical hire, the recruiting of this hire must be a strategic thrust and a top priority by senior management and human resources. If all the key technical positions have been filled, do the staff fully understand how their work supports the organization's strategy and creates value? More importantly, do they agree with the priorities in the S&T strategy? All too often, there is no full alignment of priorities among key technical staff. As part of the planning process, it is very important to flesh out these disagreements and gain commitment to the S&T strategy. One of the most important criteria to the final selection of project and program leaders is their commitment to the success of the projects they are leading.

Another area that requires a strategic review is in training and development. Too often, this is given lip service in the performance management process. When key staff feel that the organization is interested in their personal development, growth, and professional advancement, they are much more likely to give 110% effort. Commitment is a two-way street. An important strategic thrust in every organization should be a formal succession plan. Each key technical and R&D management position should be identified, and backup plans developed with the goal of having at least one ready-now candidate and one candidate developed in the next 1–2 years. A formal succession plan helps management with the development plans of current staff (i.e., to prepare them to fill future vacancies) as well as recruiting needs to fill in gaps. All too often, ambitious R&D plans get delayed or eliminated and precious funds wasted due to a single point of failure—attrition of a key technical staff member and no adequate replacement.

Finally, sufficient attention needs to be paid to the quality of the organizations' execution processes (client relationship management, proposal writing, project management, report writing, intellectual property protection, and disseminating and publishing research results). Codifying these processes and providing continuous training to key technical staff will not only improve the quality and efficiency of products and services but also improve the moral of staff and build a culture of excellence.

A Financial Strategy That Maximizes the Use of Available Funds

All organizations are forced to work with limited funding. As previously stated, the reason for developing a strategy in the first place is to determine where to focus limited financial resources. If an organization had unlimited resources, there would not be a need for strategy as it could cover the universe of options presented to it.

The organization's financial strategy for R&D spending is most often driven by external forces. For mature technology organizations, a certain percentage of its revenues is allocated to R&D. Depending on the nature of R&D, these percentages can range from 1% to 10% of revenues. For startup organizations, particularly in the biopharmaceutical industry, the financial strategy is tied to the funds needed to meet product development milestones determined by regulatory agencies such as the FDA.

Given a finite financial budget from which to work with, S&T managers must develop a strategy and allocate their funds among a portfolio of programs and projects that balance both short-term (exploitation) and long-term (exploration) needs. On the one hand, it will always be important to invest R&D dollars to extend the life of income-generating products by making them better, faster, and cheaper. However, if short-term investment is overemphasized, the future will be jeopardized,

9.10

Given a finite financial budget from which to work with, R&D managers must develop a strategy and allocate their funds among a portfolio of programs and projects that balance both short-term (exploitation) and long-term (exploration) needs.

and there will come a point where there are no new products to replace those at the end of their life cycle.

Even worse still is taking a defensive and myopic view, continuing to invest in improving your product when faced with new technology that renders it obsolete. This is often referred to as the success trap; being so enamored with current products, companies are unwilling to change even when faced with obsolescence. Bill Gates has been quoted as saying the "success is a lousy teacher." Companies become less innovative as they become more competent.

On the other hand, the future of any technology organization will ultimately depend on new product development, which can sometimes take many years to develop. In the case of Biotech products, 10- to 15-year development cycles are not unusual. So, a certain percentage of an organization's R&D budget must necessarily be spent on exploration, which requires funding projects with long time horizons. Innovation is risky and often undisciplined. Many a good idea has been abandoned before their development was truly tested. This is often referred to as the perpetual search trap.

Balancing exploitation with exploration is an art. An S&T portfolio must achieve this balance based on the current state of maturity of an organization's products, competitive pressures from competing products, and technological trends that threaten obsolescence. For further insights into the subject, I recommend reading Christianson's *The Innovator's Solution* [8] and listening to Knut Haanaes' March 31, 2016 Ted Talk on "Two reasons companies fail and how to avoid them." https://www.ted.com/talks/knut_haanaes_two_reasons_companies_fail_and_how_to_avoid_them.

A Strategy to Acquire Expensive Facilities and Equipment

Another unique aspect of S&T-driven businesses is the need for highly expensive facilities and equipment. High annual capital budgets are a way of life for most mature S&T organizations, and they often amortize capital equipment over short periods of time, 3–5 years. Smaller companies and startups have a more difficult time and often underestimate these annual costs when presenting their financial plans to funding sources.

In my experience, most capital equipment in S&T organizations is vastly underutilized. This is especially true for highly specialized equipment. Utilization rates of 20%–30% are not uncommon. While highly inefficient, these utilization rates are justified by researchers (and tolerated by research directors) by insisting that the equipment be available to them when needed at a moment's notice. There is also a culture whereby researchers insist on "owning" their own equipment as part of their department's resources. While this can be tolerated in larger organizations, it can be fatal for smaller organizations and startups with limited resources.

Two methods can be used to mitigate these costs. First, the annual cost of expensive equipment can be estimated and billed to research projects to recover those costs on an hourly or daily "use rate." In this way, the cost of the equipment can be recovered by appropriately allocating to those projects that use the equipment. By charging the cost of equipment as well as labor, the true cost of those projects can be determined. Using such a system can double and even triple the utilization rate, requiring less capital purchases and saving valuable financial resources. One word of caution however: Such an allocation scheme must be managed well so that equipment is available to scientists when needed and the costs are transparent. Otherwise, this can be a demotivating to researchers. Having used such a system, once it is installed and operating efficiently, it becomes a normal part of laboratory operations.

Another common capital equipment concern is the purchase and infrequent use of highly sophisticated equipment (e.g., electron microscopes, high-resolution mass spectrometers, and the like). A good approach is to approach colleagues at nearby universities and other R&D companies and work out a cost sharing arrangement for the use of such equipment.

It pays to allocate an administrative staff member to manage the logistics of the use rate and cost sharing systems to make it as transparent as possible to the research staff. This will go a long way to overcoming resistance to the lack of ownership and control.

Regulatory Strategy

Most science-based businesses, such as biopharmaceuticals, medical products, petroleum products, chemicals, and agricultural products are highly regulated and, in some cases, being outright banned (e.g., GMOs). Politics aside, from a scientific point of view, the precautionary principle makes a lot of sense based on age-old wisdom and folklore. Starting with Hippocrates' ancient medical principle of "First, do no harm," for centuries, our elders have preached that "an ounce of prevention is worth a pound of cure." "It is better to be safe than sorry." "Look before you leap."

Scientific progress is not linear and unfortunately tends to take two steps forward and one step back. A strong case can be made that the application of science has significantly improved the human condition from extending lifespans to feeding the world's population. On the other hand, the misapplication of science has also caused deleterious effects on mankind and the environment (e.g., thalidomide and organochlorine pesticides).

The New World Encyclopedia defines the precautionary principle as "caution in advance," "caution practiced in the context of uncertainty," or "informed prudence." The precautionary principle has three major components:

1. Anticipate harm and act to minimize potential harm.

    This is an expression of a need by decision makers to anticipate harm before it occurs and an obligation, if the level of harm may be high, for action to prevent or minimize such harm.

2. Onus of proof on proponent.

    Under the precautionary principle, it is the responsibility of an activity proponent to establish that the proposed activity will not (or is very unlikely to) result in significant harm. This is an implicit reversal of the typical onus of proof, whereby harm needs to be demonstrated.

3. Absence of scientific certainty not an obstacle.

    The precautionary principle is activated even when the absence of scientific certainty makes it difficult to predict the likelihood of harm occurring or the level of harm should it occur. There is an obligation, if the level of harm may be high, for action to prevent or minimize such harm, with the control measures increasing with both the level of possible harm and the degree of uncertainty.

It is important to note that, although this principle operates in the context of scientific uncertainty, it is generally considered by its proponents to be applicable only when, based on the best scientific advice available, there is good reason to believe that harmful effects might occur.

It is important to understand that people fear that which they don't understand. Scientific progress of late seems to be accelerating at an exponential rate and advances in genetic engineering, stem cell research, GMOs, climate change, and artificial intelligence have the potential to dramatically change the course of human evolution. This fear along with a growing mistrust of our established institutions make developing and gaining public acceptance of new products from S&T exceeding more difficult than it has ever been. Given this current climate, your regulatory strategy, and indeed your brand itself, must be beyond reproach.

> **9.11**
>
> It is important to understand that people fear that which they don't understand.

Make the Regulatory Agency Your Partner, Not the Enemy

Given the current climate of public mistrust of scientific advances, regulatory agencies have moved toward vigilant observance of the precautionary principle. It is important to understand that the agency's mission is not to prevent your product from reaching the market but to ensure that those products that they do approve or the processes that they regulate meet public safety requirements. *This is the same mission that your company has!* Thus, when an agency challenges some of your data, instead of viewing them as obstructionist, think of the agency giving you an early warning sign of what you can expect from the public if the product or process makes it to the marketplace.

I have personally been involved in supporting major pharmaceutical product development and safety evaluations requiring FDA approval. In one case, my client spent considerable time and effort working in concert with the agency to develop comprehensive preclinical and clinical protocols to address all the questions that might be raised about the products' safety and efficacy. We even spent time developing and validating new testing equipment because of some concerns that existing equipment would not adequately generate data that would address the agency's concerns. Thus, the product was approved on the first investigational new drug application (IND) and new drug application (NDA) submissions.

I have also had frustrating experiences of clients who viewed regulatory agencies as adversaries that must be kept at arm's length and whose strategy was to conduct the minimally acceptable protocols, regardless of the specific concerns that might be raised. Multiple submissions were necessary, and the time and effort to gain regulatory approval was onerous. In one case, a potential client hinted at a possible bribe by offering to pay for multiple permitting tests if we would submit only the one with the most favorable results. We respectfully declined and walked out of their office.

While these represent the two extremes, it is important to think both in the short term and long term. There is a lot of pressure brought to bear on the costs and length of time to bring a regulated product to market. This is particularly true of startup companies whose survival depends on regulatory approval before the seed money runs out. Even more reason to make the agency your ally upfront to ensure that there are no regulatory setbacks down the line.

A Regulatory Strategy Must Be Supported by Comprehensive and Unbiased Scientific Data Where the Potential Downside Risks Are Not Minimized

In developing a regulatory strategy and the scientific studies to support it, the experimental design must be comprehensive, including studies that may or may not support your product's claims of efficacy and safety. A good strategy is to try and challenge your product's efficacy and safety through rigorous testing that includes possible side effects and unintended consequences. This is not a Pollyanna approach. The earlier you identify potential problems with your product, the easier and more cost effective it will be to correct the problems or abandon the product. This approach also breeds trust with regulatory agencies if you are transparent with your experimental design. I've seen too many examples of selective studies (with the encouragement of well-intentioned regulatory consultants) designed to present products in the best possible light that fail in the long run, when inevitably, questions are raised that weren't addressed in the first place.

If your experimental design is comprehensive, inevitably you will experience some positive and negative data. There will be a strong tendency toward conformational bias that could negatively affect your interpretive reports submitted to the regulatory agency. Part of your strategy should be to challenge the data interpretation of your scientific team to develop a broader perspective of the viability of your product. While it is advisable and common practice to present your data in as positive a light as possible, it is important not to minimize the potential for adverse effects or unintended consequences that will be questioned by the regulatory agency and eventually the public.

Develop a Transparent Communications Strategy with the Public

In the long run, your product's success will not be determined by the regulatory agency but the public. As mentioned earlier, you should view the agency as your partner who could provide an early warning of potential problems before your product reaches the market. Given the degree of skepticism that the public has concerning science-based products, your regulatory strategy should be extended to direct communication with potential users. While some of the skepticism is founded on ignorance and fear, there have been mistakes that have been made in introducing new products to the market that warrant public concern.

While creating a good brand has always been an important goal, gaining the trust of the public has never been more difficult. Your communications strategy starts with a value statement that is not just a slogan but a way of life that guides your organizations' decisions about developing highly effective and safe products. In communicating with the public, you highlight both the advantages of your product and potential risks *in a way that the public can understand*. Although statistics is an important methodology to evaluate risk when dealing with the scientific community, most of the public are ignorant of statistical analysis and are fooled in the past by false claims. One of my dad's favorite expressions was "figures don't lie, but liars figure." The expression was first stated in 1889 while addressing the Convention of Commissioners of Bureau of Statistics of Labor. Carroll D. Wright was a prominent statistician employed by the US government, who said:

> The old saying is that "figures will not lie," but a new saying is "liars will figure." It is our duty, as practical statisticians, to prevent the liar from figuring; in other words, to prevent him from perverting the truth, in the interest of some theory he wishes to establish.

During the entire development process, it is wise to have a transparent communications strategy to keep the public informed. Positive data can be used to promote the potential new product. Negative data and its potential impact should be acknowledged, explained, and actions that are being taken to mitigate or eliminate the negative impacts.

The overarching question that should be asked when dealing with ambiguous data is "what can we say that will result in increasing the public's trust in our decision making".

Stakeholder Strategy

The stakeholders of your organization consist of everyone who is depending on you to deliver on your business commitments to include owners/shareholders, clients, suppliers, regulatory agencies, and the local communities in which you reside. The strategy for owners/shareholders is embedded in your vision and mission as well as expected financial returns. It is usually the overarching focus of your strategic plan. However, achieving your owner/stakeholder strategy is highly dependent on success of the five strategies in your strategic agenda. We have already discussed the importance of regulatory agencies and will now focus on a strategy for the remaining stakeholders; clients, suppliers, and the community in which you operate.

Client Strategy

In the section on client feedback, we emphasized the importance of gaining intelligence from existing clients on their current problems and future opportunities and the value you bring to their organization. Armed with this intelligence, you can then develop a client strategy to accomplish two important goals: improve the quality and responsiveness of your current products and services and anticipate your client's future needs for new products and services. The first client feedback question should provide you with the intelligence you need to formulate the first goal of your client strategy. "How would you rate the quality, responsiveness, and value of the products/services that we have provided over the past year on a scale of 1–10? How does this compare to our competition?" The scores on quality will assess your product, scores on responsiveness will assess your processes, and scores on value will evaluate the justification for your costs. These measures will tell you how well you are currently serving your clients. A key follow-up question that I have found to be extremely useful is to ask what your organization would specifically have to do to improve your scores. This is a method of extracting valuable information even when the client has given you a high rating and will form the basis of your strategy going forward to improve the quality and responsiveness of your current products and services.

To anticipate your client's future needs for new products and services, the answer to the following question will help. "Where do you see your market going in the next few years and the products/services you will be needing from us?" This is an open-ended question that probes clients about the future and creates a dialog where you can introduce new ideas. This can be very helpful to include your client's needs in generating your research agenda in the strategic plan.

As previously discussed, your goal should be to become a strategic partner with your clients, involved in their product development strategy and marketing plans. A key question to have answered is "What level of impact is our research having on your organization's strategy? What more can we be doing to become a strategic partner?" This can be presumptuous with a new client, but in time, as you gain more trust by your actions, several of your key clients will better appreciate the valuable role you can play by being in their inner circle. They will begin to trust you well enough to share the more difficult problems that they are facing or opportunities that they wish to pursue. This should be the goal of your client strategy.

Supplier Strategy

Once you have internalized the process of client feedback and implemented a client strategy, you will understand the value of developing strategic suppliers. It is simply a role reversal, where you are now the client and appreciate the value that accrues from screening and evaluating vendors and moving them up the value chain for your organization to preferred providers and ultimately to strategic partners. This is a process that can take many years with a low percentage of success. It is therefore important to have a supplier strategy to get you there. It starts with setting high standards that you expect them to meet and providing feedback on their performance. The same feedback questions that you developed for your clients can be used if the supplier does not have his own.

This approach is especially useful in the biotechnology business, where a great deal of startup companies depend on a virtual business model that rely on CROs to generate their preclinical, manufacturing, and clinical research studies. Evaluating the performance of several suppliers, downselecting a few that are trustworthy, and then sharing your development strategy with them is a valuable way to develop a stronger commitment from suppliers and improve the quality and timelines of your research. I have experienced numerous examples where my staff developed ownership and a deep commitment to a client's preclinical research program because they felt a part of the client's research team. I have also experienced the opposite effect, where clients have taken a handsoff approach to contracting out testing protocols, resulting in just a business transaction and less enthusiasm on both sides. Trust is a two-way street.

Community Strategy

An often-overlooked part of an organization's strategy is the position and actions it takes to become a good corporate citizen and build social trust. The community in which you operate can, and often does, have a direct impact on your license to operate and the permits require for continued operation. In addition, your local reputation is an important part of your brand, which can affect your business wherever you sell.

Rather than taking a passive approach and reacting to community requests from time to time, a good strategy is to decide in advance what your contribution to the community is going to be and communicate it widely to the community leaders and the public at large. You can then concentrate your resources (pro bono contributions and money) in a deliberate way to maximize their impact.

As an example, our laboratory was in a small town with a strong reputation for academic excellence. Being a science-based organization, we decided to contribute to the town by establishing a local science fair for K-12 students. We organized and funded the fair, and our staff along with school teachers evaluated the science projects submitted. Awards were given to the top projects in each grade level. While it took a couple of years to gain traction, our laboratory gained positive recognition in contributing to the community. This strategy worked better for us than just responding to the numerous requests for community donations. In establishing a community strategy, strive to focus your attention and resources in an area where you can make a difference.

Your final strategic agenda should include between five and ten strategic thrusts focused on one or more of the strategies mentioned earlier. In some cases, you may decide that the critical success factors for your organization involve only some of the above five strategies and that you need to focus on just two or three. Don't make the mistake of pursuing too many strategic thrusts that require substantial commitments and mind share of your leaders. Making these kinds of decisions based on your data analysis will help you allocate your resources where they will have the biggest impact. Once again, strategy is about making tough decisions about where to invest your limited resources.

☐ Chapter Summary

The strategic planning process starts with developing or updating the organization's mission, vision, and values and finishes with a strategy designed to fulfill its mission and achieve its vision.

The strategic planning process consists of seven steps:

Self-assessment

> Product/service—Evaluate the quality and competitiveness of your product

> Processes—Evaluate each of your critical processes to be sure that they contribute rather than subtract from your products' value proposition

> People—The most important evaluation is of the people in your organization that make a difference, i.e., your leadership

Client Feedback

> Obtain valuable client feedback from a select subset of your client base that is most important to your current and future business success

Competitor Analysis

> Product/service—What are the specific reasons why certain customers purchase a competitor's product/service?

> Processes—Do competitors produce their product faster, cheaper, and of a higher quality than you do?

> People—It is a good practice to get to know your competitor's key leadership as well as you know your own

Technology Trends

> Incremental—The relative rate at which both you and your competitors are improving your existing products and services (i.e., making them better, faster, and cheaper)

Evolutionary—The development of a new technology or marketing business model that results in an immediate and substantial competitive advantage

Revolutionary—Seemingly irrelevant emerging technologies can be innovatively applied to render your product, service, and even your company obsolete

Market Analysis

Primary data—Gathered from interviews with industry participants and observers

Secondary data—Gathered from published sources

Stakeholder Requirements

Regulatory—Understanding specific regulatory requirements will likely be the most cost effective and fastest route to product approval and the development of trust among the regulators about your company's commitment to safe and efficacious products

Employees and community—Understanding the needs of your employees and community and the concept of corporate social responsibility

Strategy Synthesis

Clearly define the organization's mission (why do we exist?), vision (what do we want to become?) as well as values (how should we behave?) that will ultimately determine its culture

Market Strategy

A winning market strategy is one that allows you to successfully compete in a favorable, highly attractive market segment with a subset of clients who highly value your products/services

Product Strategy

Develop an overarching technology strategy

Develop a portfolio strategy to support the technology strategy

Select the most promising projects in your portfolio

Resource Strategy

Staffing strategy—Identify and deploy key technical staff to execute the research portfolio

Financial strategy—maximizing the use of available funds

A strategy to acquire expensive facilities and equipment

Regulatory Strategy

Judicious deployment of the precautionary principle

Building social trust

Stakeholder Strategy

Client strategy—Improve the quality and responsiveness of your current products and services anticipate your client's future needs for new products and services

Supplier strategy—Appreciate the value that accrues from screening and evaluating vendors and moving them up the value chain

Community strategy—Take action to become a good corporate citizen and build social trust

☐ **References**

1. Matheson, David, Matheson, Jim, *The Smart Organization, Creating Value through Strategic R&D*, Harvard Business School Press, Boston, MA, 1998.
2. Peters, Thomas J., Waterman, Robert H., *In Search of Excellence—Lesson's from America's Best-Run Companies*, Harper & Row Publishers, New York, 1982.
3. Diamandis, Peter, Kotler, Steven, *Abundance, the Future Is Better than You Think*, Free Press, Simon and Schuster, New York, 2012.
4. Porter, Michael, *Competitive Strategy, Techniques for Analyzing Industries and Competitors*, The Free Press, Simon and Schuster, New York, 1998.
5. Robert S. Kaplan, Norton, David P., Using the balanced scorecard as a strategic management system, *Harvard Business Review*, 1996.
6. Kim, Chan W., Mauborgne, Renee, Blue ocean strategy, *Harvard Business Review*, 2005.
7. Schmitt, Bernd H., *Big Think Strategy, How to Leverage Bold Ideas and Leave Small Thinking Behind*, Harvard Business School Press, Boston, MA, 2007.
8. Christensen, Clayton, *The Innovator's Solution*, Harvard Business School Press, Boston, MA, 2003.

Managing the Execution
Translating Your Strategic Agenda into Actionable Objectives and Managing to Achieve Those Objectives

Once a compelling strategy is formulated and a strategic agenda developed consisting of several strategic thrusts, the second process of the Performance Trilogy® involves execution. The execution process involves translating your strategic agenda into actionable objectives and managing to achieve those objectives. This is easier said than done. Nearly 80% of all strategies fail primarily due to execution [1]. A metaphor that I find useful is to think of strategy as potential energy and execution as kinetic energy. Strategy is just an idea. Execution is when you actually take action to make things happen. As mentioned previously, Peter Drucker points out that knowledge, wisdom, and expertise are useless without action. Managing execution takes hard work, sweat, and practice. Strategic plans are just good intentions until followed up with hard work and determination (Figure 10.1).

In my experience, I have rarely seen this translation from strategy to execution done very well for three reasons. First, translating strategic thrusts into actionable objectives requires that you pinpoint the critical success factors that ensure the desired outcomes. There must be a high degree of confidence that if you achieve the designated objectives, the desired outcome is assured.

Rarely have I seen this level of alignment. In the quest to produce actionable objectives, managers often take the easy way out with objectives that are easy to quantify and achieve. Oftentimes, the critical success factor in achieving a strategic thrust involves betting on a highly specific objective that is very hard to achieve. For example, the winning of a strategic contract that would establish market leadership, developing that unique product that transforms the basis for your business model, or making a strategic hire that changes the competitive landscape. Committing to this make or break objective that assures meeting your strategic thrust takes considerable courage. Most managers feel safer hedging their bets with safer objectives even if they don't assure success.

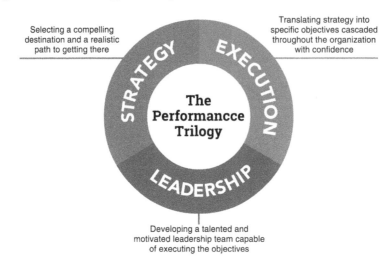

Selecting a compelling destination and a realistic path to getting there

Translating strategy into specific objectives cascaded throughout the organization with confidence

Developing a talented and motivated leadership team capable of executing the objectives

FIGURE 10.1 The Performance Trilogy.

I am reminded of the story of a man who was walking down a city street at night and noticed an older gentleman stooped over the curb under a lamppost. When asked what he was looking for, the old gentleman stated that he lost his keys in the alley and was looking for them. The man asked the obvious question "If you lost your keys in the alley, why are you looking for them out here"? The old timer said, "because this is where the light is." As silly as this sounds, I have seen many examples of highly intelligent managers pursuing objectives that, even if achieved, would not ensure the success of the targeted strategic thrust.

Second, individual team members have personal aspirations and agendas that do not align 100% with the proposed organizational strategy and tend to go off on tangents. I use the term institutional objectives (IOs) to describe the high-level objectives that drive the organization's strategic agenda. These IOs become the performance plan of the CEO or COO of the company and his immediate management team.

To implement these IOs, they must be broken down into more discrete elements and cascaded to the next level of management to implement. The larger the organization, the more cascading is needed, and at every level, the original IO gets diluted due to the personal agendas of each level of management.

The perfect execution plan is one where every employee comes into work each morning, knowing how his performance will contribute to meeting one or more IOs. This is rarely the case but the closer an organization comes to this ideal, the more likely it will succeed in executing its strategy. Therefore, the first questions I ask of employees when evaluating the performance of an organization are "What is the strategy of your organization?" "Do you agree with it?" What role are you playing in helping to execute that strategy?" I will be presenting below a disciplined performance management process that mitigates this problem and helps ensure that there is a high alignment of each employee in the organization.

Third, senior leaders incorrectly assume that the translation of strategy and development

10.1

The perfect execution plan is one where every employee comes into work each morning, knowing how his performance will contribute to meeting one or more IOs.

of a performance management process is not their responsibility because it deals with operations. They believe that they shouldn't be "getting their hands dirty" and delegate execution and the performance management process to their subordinates and middle managers. Bossidy, in his book on *Execution* states "Many people regard execution as detail work that's beneath the dignity of a business leader. That's wrong. To the contrary, it is a leader's most important job.

People think of execution as the tactical side of business, something leaders delegate while they focus on perceived 'bigger' issues" [2]. Execution is not just tactics, it is a discipline and a system. It has to be built into a company's strategy, its goals, and its culture. And the leader must be deeply engaged in it.

As a result of this mindset, the time and effort of leadership teams throughout the organization, the scarcest resource in most organizations, are not being fully used to advance those critical IOs that would ensure success. The hard work that Drucker talks about and exemplified in well-managed organizations is the attention and persistence directed at ensuring that everyone is working toward achieving the IOs [3].

Many a good strategy has failed from poor execution due to senior managers who "abandon" their direct reports once performance objectives are developed. It is the responsibility of the senior manager and architect of the strategy to ensure not only that the strategy gets translated into actionable objectives but also that performance is managed. This is accomplished not by macromanaging or micromanaging but by active management and regular governance on the progress of objectives. This involves supporting each team member with removing organizational barriers, identifying performance shortfalls, determining lessons learned, and correcting deficiencies.

This concept of "managers must manage; management is an active verb" struck home to me when I read Geneen's book on *Management* [4]. His description of what it takes to truly run a business sticks with me to this day. The best way to run a business with the best hope of eventual success is to do it as you would cook on a wood-burning stove. How do you cook on such a primitive stove? Because you know that you cannot control all the elements of fire, wood, air flow, etc., you keep your eye on everything at all times. You follow the recipe to an extent, but you also add something extra of your own. You do not measure out every spice and condiment. You sprinkle here, you pour there. And then you watch it cook. *You keep your eye on the pot.* You look at it and check it from time to time. You sniff it. You dip your finger in and taste it. Perhaps you add a little something extra to suit your own taste. You let it brew a while and then you taste it again. And again. If something is wrong, you correct it. Whatever you do, the most important thing is to keep your eye on it. You don't want it to be ruined when you are off doing something else. When it is done to your satisfaction, you're right there to take it off the stove. In the end, you will have a pot roast or lamb stew that is the very best you could possibly make, a joy to your palate and a tribute to your ability as a cook. It will taste far better than any slab of meat you cook automatically by pushing buttons on a microwave oven. That is how you would cook on a wood-burning stove when nothing is preset. And that is the frame of mind to take into the art of conducting and building a successful business.

Both the development of the strategy and translation of it into an execution plan with actionable objectives must be done jointly between you and your team members. As discussed previously on leading the Performance Trilogy, obtaining not only buy-in but also enthusiasm from team members is critical to success. Objectives should simultaneously support the strategy as well as the personal goals of every team leader. This is a good definition of "alignment."

Each of the team member's objectives must be directly linked to the advancement of the strategy such that the output of each objective results in the desired strategic outcome. This requires practice and experience to do well. The minimum requirement should be at least a 70% alignment (i.e., 70% of the objective's output contributes to the strategic outcome). Any percentage lower than that would require a serious reexamination of that team member's suitability for and/or willingness to take on the assignment. As the manager of the process you not only need to ensure such alignment but also make sure that it is happening at each level of the organization. Alignment provides a way of linking strategy and people and integrating them with customers and process improvement [5].

Unfortunately, too many managers view setting performance objectives as a necessary evil and an undesirable chore. After reading this book, I hope that it will increase the awareness of the importance of performance planning and monitoring as a fundamental element in the execution process.

☐ Ensuring That the Strategy Is Rigorously Translated into Performance Objectives That Are Actively Managed throughout the Organization

A disciplined performance management process is essential to ensure that key elements of the strategic plan are being properly implemented. I have described a codified performance management process later based on proven fundamentals to support the execution of your strategic plan. While not original or foolproof, it does provide specific and detailed guidance on actively managing (not macro- or micromanaging) your team to ensure that the performance objectives that directly support meeting your strategic agenda are receiving top priority. It is a systematic process for monitoring, evaluating, ensuring, and communicating the performance of each organizational unit of your organization. Properly executing each of these steps is critical to success but is easier said than done. Each step has its share of pitfalls that doom many strategies, which is why over 70% of all strategies don't meet expectations. We will point out some of the pitfalls to avoid in the later sections. The performance management process consists of five distinct steps outlined in Figure 10.2.

Performance Management Process

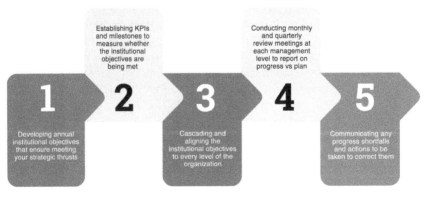

FIGURE 10.2 The performance management process.

☐ Developing Annual IOs That Ensure Meeting Your Strategic Thrusts

The first step in the performance management process is to list your strategic thrusts in your market, product, resource, regulatory, and stakeholder strategies and brainstorm the critical success factors needed for each thrust. You can

10.2

Never mistake activity for progress.

then develop specific actionable objectives for each strategic thrust that *must* be met to achieve success. As mentioned previously, this is not an easy exercise. There are many objectives that would contribute to each of the strategic thrusts but wouldn't guarantee success. You must identify and focus on the *critical few* that ensure the success of each thrust area otherwise, you will get bogged down in too many objectives and deceive yourself that you are making more progress than you are. Never mistake activity for progress.

As a way of illustration, during a strategic planning exercise for one of our clients, we conducted a customer survey and identified a general dissatisfaction with level of engagement and service being provided by our client's technical project managers. This was viewed as a serious threat not only to the current business but also future business growth. As a result, we identified "becoming more customer focused" as an important strategic thrust in the strategic planning process. After considerable brainstorming, we determined that

- The organization had hundreds of clients every year that were all treated exactly the same

- Several of the clients were mission critical while many were opportunistic

- Project managers were overwhelmed by the number of projects managed

- Clients were treated as funding sources for internally generated ideas with little consideration for client needs

- Project managers from different departments were unaware of projects being conducted from other parts of the organization

- There was no systematic external feedback on project performance

Does any of this sound familiar? Despite words to the contrary in company brochures and advertisement, it is rare to see a science-based organization that is really customer focused.

We determined that the critical success factor for this strategic thrust, becoming more "customer focused" were to identify the mission critical client base; develop and implement a client engagement process for that client base; dedicate management time and resources; and measure improvement through continuous client feedback. For each research department, IOs included

- Develop meaningful evaluation criteria to downselect mission critical or "key" clients

- Establish a key client list based on the evaluation criteria and communicate it throughout the organization

- Conduct an annual client visit by senior management to listen to client needs and concerns

- Develop a client strategy based on feedback from the annual client engagement

- Assign a key client relationship manager to continuously manage the client relationship and obtain monthly client feedback

- Develop a client satisfaction survey and conduct the survey after every project

- Document "lessons learned" from every project

- Conduct follow-up meetings with clients on areas of concern

Although these IOs would require a major amount of time, they were essential to ensure that the strategic thrust would have a successful outcome for selected, mission critical customers.

The feedback obtained from key clients contained an operational element; how well are we performing currently, as well as a strategic element; what could we be doing to support you in the future? Every organization faces both operational and strategic issues during its strategic planning.

Operational performance management focuses on meeting current commitments to owners/stockholders, clients, and stakeholders as well as improving operational efficiency, effectiveness, and timeliness. The goal is to maintain and grow the current business and is present-oriented with IOs measured daily and monthly.

Strategic performance management focuses on strategic objectives that build the organization's future and achieve its vision listed in the strategic plan by developing new products and expanding into new markets, thereby growing in size, reputation, and commercial income. By its nature, strategic performance management is future-oriented with IOs that are measured over several years.

Another example of a strategic thrust that was developed for a science and technology (S&T)-based consulting and contract research organization was to "become the employer of choice for scientists and engineers." We recognized that a critical success factor in growing an S&T-based business was to attract, recruit, and hire the best and the brightest technical talent in their respective fields who could provide the thought leadership and technical skills to compete in a highly competitive consulting market.

We brainstormed the critical success factors that would aid in our recruiting efforts and determined that to be the best place for scientists and engineers to work we would need to start from within. When asked what is it like to work in our organization, we wanted our senior technical staff to give a glowing recommendation. In a tight-knit technical community, when being recruited, we believed that input from colleagues plays an important role in the decision-making process. We surveyed our senior staff and not surprisingly came up with the following suggestions for the ideal workplace:

- Challenging technical assignments and difficult problems to solve

- A high degree of autonomy and lack of bureaucracy

- A reward and recognition system based on merit

- Supportive management team that listened and learned

- A collegial atmosphere and fun place to work

We felt that these suggestions were a good place to start and needed to translate them into actionable IOs. Once again, we focused not on easy objectives, but ones that we felt would be critical to ensuring success at become the best place for scientists and engineers to work.

- Established bidding criteria that eliminated projects of a routine nature and focus on challenging technical assignments

- Developed a three-stage career ladder of equal stature and compensation that afforded options for technical staff in technical leadership, project management leadership, and general management leadership.

- Implemented a reward system that was merit based and emphasized the ability to sell technical projects, manage projects, or manage people.

- Conduct an annual staff satisfaction survey and develop improvement goals for any survey item lower than 8 out of 10.

- Establish an annual reward and recognition program (our version of the Oscars) overseen by peers and including multiple categories.

- Hold an annual company Oscar party attended by spouses in recognition of outstanding performance.

We were highly successful in recruiting top talent to our organization we believe due to the steps we took to achieve the critical success factor for becoming the employer of choice in our field. An important lesson learned was to recognize that the most important factors were the type of work assignments that challenged our senior technical staff and the degree of independence they had to execute their project work, including budget control. Next, the reward system and compensation needed to be merit based and widely recognized as fair. And finally, the parties and trophies that publicly recognized accomplishments were "icing on the cake." Napoleon once said that "the most important lesson I learned in military school was that men are willing to die for medals."

☐ Establishing Key Performance Indicators and Milestones to Measure whether the IOs Are Being Met

There is a widely accepted management principle worth subscribing to. "What gets measured gets managed." It is important for any strategic thrust to know your current level of performance, what performance you aspire to, and how will you know if you get there. This can only be assured if you develop performance indicators that you can measure for each IO. While there are many performance indicators that can be used for each IO, there are just a few that ultimately capture the essence of whether you have achieved your objective. These are called key performance indicators (KPIs).

In an earlier example on getting more customer focused, several in-process performance indicators can and should be measured, such as whether client strategies were developed, client relationship managers assigned, and whether client meetings have taken place. However, the KPIs are the cumulative scores on the client satisfaction surveys and the annual client feedback report. On a scale from 1 to 10, the initial satisfaction scores for each research department ranged from 4.3 to 6.5. The KPI for client satisfaction for all departments was set at 8, with the expectation that it would be raised to 9.

Using client satisfaction scores as KPIs can reap great rewards if performed well and used properly. In my experience, most science-based organizations either do not have such a KPI, or if they do, their execution actually hurts rather than helps their relationship with clients. In designing and executing a client feedback process, there must be a genuine management commitment to commit the time and resources to obtain maximum feedback and act on the information received quickly and responsively. Nothing hurts a client relationship more than asking for feedback, obtaining a negative response, and not following up immediately.

First and foremost, the survey must be designed to ask the right questions that target both current performance and expected future performance. As mentioned previously, this will provide information on how well you are providing current products or services and what products or services may be required in the future. This can be used to drive your S&T agenda. Second, the client feedback process should be selectively used on your most important clients, since to do it right requires a significant amount of time and effort. If you have a client relationship manager, formal or not, that person should conduct the survey in person if possible. If not, then by phone or videoconference. We are all familiar with impersonal email surveys and how ineffective they are. The average response rate is usually less than 10%. By using direct contact and persistent follow up, you should aim for greater than 80% response rate from your key clients.

These survey results, both individual and cumulative, should become part of the performance reporting system that gets published and discussed at each management meeting. Any negative feedback from surveys should be immediately communicated to the S&T team working on the project and the cognizant manager. The appropriate manager then needs to follow up expeditiously, expressing concern and a willingness to correct the problem.

While time consuming, the client feedback process is an excellent way to ensure that your entire team is client focused and committed to meeting your client's current and future expectations. It is also an excellent method to objectively measure an S&T team's performance on individual projects as well as cumulative annual performance. I have heard many excuses for not implementing such a feedback system. "We don't have the time." "We don't want to bother the client." "Our clients are reluctant to give negative feedback." Once an organization successfully implements such a system and sees how effective it can be, they will wonder why they didn't implement it sooner.

Let me give you an example from personal experience. After implementing such a system, I met with a division director from one of our government clients. At the time, my research staff were conducting several projects supervised by the division director's technical project managers. When asked how many staff worked for me, I responded, "my staff don't work for me they work for you." He laughed and responded' "all my technical contractors say that." I responded, "that may be true, but I can prove it." This got his attention. I showed him a table of performance scores given to my S&T teams by his technical project managers for each of the projects conducted during the year. I then showed him the criteria I used to conduct annual performance reviews for my research staff. The criteria were heavily weighted toward these client feedback scores.

I then said "the annual raises and promotions given to my staff are based on their performance reviews. The performance review scores are heavily dependent on the feedback scores that your technical project managers give my staff. Therefore, your project managers determine the raises and promotions received by my staff. Who do you think my staff work for, me or you?" The next time we had to compete for a contract with this government client, we scored a perfect 10 on the management section of our proposal.

☐ Cascading and Aligning the IOs to Every Level of the Organization

If you have successfully completed Steps 1 and 2 of the performance management process, you and your leadership team should have a short list of goals (i.e., IOs) that *must* be achieved and feel confident that, if accomplished, you will be successful in achieving your strategic thrusts. When your organization is small and your direct reports are the actual implementers, it is much easier to ensure that progress is being made and no one is off on any unproductive tangents. When I managed a laboratory of 75 staff and had five direct reports, I remember feeling very confident at the end of every planning meeting that the goals that were set and the KPIs to measure them were well understood by my direct reports and that it was easy to monitor performance. When my organization consisted of over 500 staff, it became a lot harder and required much more work to ensure that managers down the chain of command were aligned with the strategic agenda.

When the organization becomes much larger, several additional layers of management are created. This makes it much more difficult to ensure that the IOs given to the top management team are being properly executed at the lower levels of the organization. Most of us have participated in the parlor game where you pass on a specific piece of information to the person next to you and ask that they pass it on. After passing through several people, by the time it gets back to you, the original information is hardly recognizable! This is quite normal like the deterioration of a recording or disc after it has been played many times or errors in DNA after having replicated numerous times.

This is why it is important to pay close attention to Step 3 in the process. The larger the organization, the easier it is for lower level managers and individuals to misinterpret, or in some cases, ignore the institutional goals and pursue their personal agendas. The ultimate goal should be that every employee is committed and focused on translating and adapting the institutional goals to their job functions every day that they come into work. Remember that if the institutional goals are met, it will ensure the success of the organization's strategy. It's that important. As I have mentioned previously, a good rule of thumb is that 70% of all employees should be at least 70% aligned with the IOs. Where do you think your organizations stands on this KPI?

Some useful practices to follow to ensure this level of alignment are as follows:

- The number of IOs should be small.

    Too many managerial leaders feel a need to overhaul their entire organizations during a change initiative. This requires the establishment of too many goals. When there are too many priorities in a strategic plan, there are no priorities at all. It becomes nearly impossible to put the time and energy into achieving more than about five organizational goals in any given year.

- The objectives should be crystal clear.

    As demonstrated by the parlor game above, transmitting information through multiple layers of an organization is fraught with difficulty. The simpler the IO is to describe and the smarter the KPI is, the higher degree of success that can be achieved. A common acronym SMART is often used to describe an objective: Specific, Measurable, Attainable, Relevant, and Time bound.

- You should review the objectives of each of your manager's direct reports

    As I have mentioned, if the organization is small, you can feel confident that your direct reports are focused on implementing the institutional goals and can monitor their execution directly. As the organization gets larger and multiple layers of management are created, it becomes more and more difficult to ensure that the appropriate objectives have been cascaded down the organization and properly translated by each manager. In other words, how can you be sure that every employee's objectives are directly related to supporting the company's IOs?

    A useful approach is to adopt a "one over one" approval system. Your direct reports sign off and approve their staff's performance goals and you review them. This small increase in effort allows you to ensure that there is good alignment at least two management layers below you. By reviewing the performance goals of your direct report's staff and discussing any misalignment issues, you can coach your direct reports of the importance of focusing on the KPIs of the institutional goals. Your direct reports can then repeat this process two layers below them. In this way, you have built in a good quality control check at each management level. Remember the ultimate goal is to have every employee come to work each morning, knowing that what they are working on is contributing to the organization's strategic agenda.

☐ Conducting Monthly and Quarterly Review Meetings at Each Management Level to Report on Progress versus Plan

In Steps 1–3 of the performance management process, institutional performance objectives and KPIs are established for the senior leadership team and each of their direct reports. The objectives are then cascaded to the program, division, and department levels of the organization. These objectives represent the performance goals for each of the managers of the organization and should be made transparent to the organization. Many of these objectives are objectives of the team or organizational unit that will require teamwork and group effort to meet. A key success factor in achieving these objectives will be developing team ownership and shared responsibilities.

After all of this work, you now have a plan. This is not the time to become complacent! A plan is nothing more than a wish list for each of your managers to follow. Now comes the hard part where the rubber meets the road; the execution of the plan by each of the responsible managers in the organization. This is the point where you take off your "leaders" hat and put on your "managers" hat; making sure that everyone in the organization is pursuing the company's performance objectives. You will simultaneously be putting on your "coach's" hat, ensuring that you are supporting each staff member in achieving their personal goals. Once again, you should be shooting for at least 70% alignment between organizational and personal goals.

The primary process for monitoring both organizational performance and development of staff is through management meetings. I will never forget the day that my daughter accompanied me on a "take your daughter to work" day. When asked by my wife when she got home "What does Daddy do all day?" She answered, "he spent all day in meetings!" There is no doubt that the primary communication mechanism for monitoring performance, evaluating progress on operational and strategic goals, and problem solving is the infamous personal and group management meeting. Meetings are to managers what hammers and nails are to carpenters; the tools of the trade. It is the primary mechanism of managing individuals, groups, and organizations. If you want to be a good manager, you will need to conduct effective meetings. I have always been amazed at how few managers really understand how to conduct effective meetings. This is a fundamental worth mastering.

10.3

If you want to be a good manager, you will need to conduct effective meetings.

Given the amount of time spent in management meetings and their importance, having a codified process for these meetings seems prudent. Your first impression will be that it appears like a lot of work. In the beginning, there will be a learning curve for sure. But once you get good at meeting management, you will find that it saves you more time than you spend preparing for them. Remember that meetings are the software of the organization and need to be managed well for the organization to function [2].

I have observed hundreds of management meetings in several different organizations and found that many of them were not focused on the important priorities of the organization (i.e., institutional goals), were very inefficient, and for the most part dominated by the meeting manager. There was a general lack of planning and very little documentation or follow-up. It's no wonder that most staff dread management meetings!

While there are many reasons for conducting different meetings in an organization, I have described a codified approach later to specifically conduct monthly and quarterly performance management meetings to assure that performance objectives are being met, which is Step 4 in the performance management process. If followed closely, these meetings will improve transparency, foster teamwork, and support problem solving in an efficient manner. And yes, they can be fun as well.

Monthly and quarterly performance management meetings should be well planned and conducted by each unit manager in the organization with appropriate follow-up. At each meeting, progress toward meeting the IOs are reviewed, shortfalls are discussed, and corrective actions are planned. Minutes of the performance management meetings are generated to communicate progress and action plans to correct problems. The role of the manager should not be to dominate the meeting but to facilitate discussion, support problem solving, and prevent the meeting from veering off into nonproductive tangents.

There is a time and a place for socializing, brainstorming, debating, and other important activities, but the performance management meeting is not one of them.

In the performance management process, IOs and KPIs have been divided into two segments, strategic performance management and operational performance management.

Strategic performance management focuses on strategic objectives that build the organization's future and achieve its vision such as the any anticipated major research accomplishments listed in the strategic plan as well as growth in size, reputation, and commercial income. By its nature, strategic performance management is future-oriented with IOs that are measured over several years rather than monthly or annually. Despite the fact that strategic objectives are very important, they may not be perceived as urgent and may tend to be overshadowed by the more immediate day-to-day operational performance objectives. The strategic performance management process has been designed to overcome the tendency to ignore lack of progress with strategic objectives. This is accomplished by scheduling regular quarterly meetings focused exclusively on longer-term strategic objectives.

Operational performance management focuses on meeting current commitments to clients, customers, and stakeholders as well as improving operational efficiency, effectiveness, and timeliness. As such, meeting operational objectives are both important and more immediate and are often a major component of year-end performance appraisal scores. Therefore, the operational performance management process involves regularly scheduled monthly meetings at each management level focused on early detection and communication of potential problems that might impact meeting the annual operational objectives.

An example of the key elements of the performance management meetings developed for one of my clients is shown in Table 10.1. The KPI scorecards are specific to each S&T organization, but this is a good starting point for most organizations. While it doesn't guarantee that the appropriate management actions will be taken to ensure performance, there is sufficient transparency to allow for supervisory intervention if necessary.

Strategic Performance Management

The strategic performance management process starts at the senior management level by developing annual IOs that support the strategic thrusts listed in the strategic plan and establishing KPIs to measure whether the objectives are being met. The annual IOs are then cascaded down (delegated) to each level of management. It is the responsibility of each manager to monitor the performance of each of his direct reports and take appropriate steps to ensure that the IOs are met. As a guide to this process, a series of management meetings along with supporting tools have been developed to review the organizational strategic performance versus plan and communicate progress.

Strategic Review Meetings

Rationale

The final responsibility for delivering the expected performance of the organization resides with the senior leadership team. The leadership team's objectives incorporate the

TABLE 10.1 Performance Management Meetings

	Strategic Management Meetings	Operational Management Meetings
Focus	• Strategy and longer-term objectives • Anticipated major achievements listed in strategic plan • Multiyear initiatives that build the organization's future such as growth in size, reputation, and commercial income	• Tactics and short term, annual objectives. • Meeting current yearly commitments to all stakeholders • Improving operational efficiency, effectiveness, and timeliness
Schedule	Quarterly	Monthly
KPI scorecards	• Progress on anticipated major achievements • Quarterly client satisfaction and impact scores • Progress on improving customer satisfaction scores • Growth in regional/ international Clients • Researchers with international reputations • Publications/ "active" partnerships • Researchers • Commercialization deal flow/intellectual property (IP) • Portfolio development	• Revenue targets met • On-time project delivery • Product quality scores • Proposal pipeline • Client satisfaction scores (Project feedback/summaries) • Staff utilization • New hires/turnover • Coaching/training initiatives • Client funding • Program progress update
Outcomes	• Action plan generated at each meeting by manager • Meeting minutes distributed one level up and one level down	
Monitoring of progress	• Performance review template used to monitor progress versus plan using traffic-light approach (Green—on plan; yellow—behind plan with corrective actions in place; red—seriously behind plan with risk of not achieving IOs in the proposed timeframe.	

institutional outcomes expected of the organization. These objectives are then cascaded to each management level in the organization. The organization's success in meeting these objectives is dependent on each manager delivering on their objectives as measured by the corresponding KPIs. Quarterly institutional strategic review meetings have been planned that allow managers to report on their progress versus plan in meeting their IOs.

Schedule

The head of the organization (e.g., CEO, managing director, director general) will conduct quarterly strategic review meetings with his leadership team. The CEO meetings should be scheduled quarterly and be allowed sufficient time for the generation and distribution in advance of the strategic performance scorecards (see meeting tools later). The leadership team should meet their respective management teams and consolidate their teams' input before the CEO meeting. This process should take place throughout the organization.

Attendees

At the strategic review meetings, in addition to the cognizant manager and his leadership team, key technical and administrative staff should be invited. Since these meetings take place quarterly, this doesn't represent a significant number of overall man hours out of their schedule and can reap significant benefits. These include improved communication and transparency, increased ownership in the plan by key staff, and valuable feedback and problem solving of the plans' shortfalls.

Purpose

The primary purpose of the strategic review meeting is to review the progress toward meeting the *long-term* IOs identified in the strategic plan. The focus is entirely on the strategic goals. The tendency to revert back to day-to-day operational issues should be strictly avoided. In addition to reviewing progress on the organization's anticipated accomplishments and performance objectives, these quarterly meetings should be used to identify market and technological developments that potentially affect the organizations' strategy. They should also review progress on the organization's long-term leadership development plan.

Preparation and Inputs

At least 2 weeks before each of the strategic review meetings, the strategic performance management scorecards are distributed (see meeting tools later). After reviewing their scorecards and meeting the leaders in their organizational units, each member of the leadership team should prepare a consolidated presentation for the cognizant manager, focusing primarily on the KPI shortfalls, reasons proposed for the shortfall, and specific actions to be taken in the next quarter to improve the KPI.

Agenda

The strategic planning meeting should last for about 6–8 hours depending on the size of the leadership team.

Management introduction and overview—30 minutes—The cognizant manager at each level of the organization will introduce the meeting and update his leadership team on significant external events. Based on the manager's wide network and extensive travel (e.g., board of directors' meetings, attendance at a higher-level performance management meeting, international meetings, discussions with key clients and alliance partners), he can communicate the latest strategically relevant market and technological developments to the team as well as the outcomes from the strategic review meeting one level higher. He will also present an overview of the unit's strategic management's scorecard and point out areas that need attention (traffic light approach described in Table 10.1).

Leadership team members with strategic objectives—30 minutes each—Each member of the leadership team will report on the progress versus plan of their strategic objectives using the performance review template (see later). The focus will be on identifying the red flags, causes for the lack of progress, actions taken in past quarters (see action log later). After each presentation, 30 minutes should be allocated for discussion, to include questions and suggestions. This is an opportunity for the leader to obtain the collective insights of the team in problem solving any barriers to meeting objectives.

Special topic/invited speaker—30 minutes—Since the focus of the meeting is strategic, time should be given to a special topic for discussion that will stimulate innovative thinking and reinforce the importance of the strategy. This could include the presentation by a staff member to recognize him for an outstanding achievement that advances the strategic agenda. Alternatively, the cognizant manager could invite a distinguished speaker to present on a topic relevant to the organization's strategy.

Brainstorming—60 minutes—There should be an open session where any attendee can comment on the presentations given, ask questions, and provide advice on overcoming shortfalls in the performance objectives.

Wrap up—30 minutes—The cognizant manager should summarize the results of the meeting and highlight the key actions that need to be taken in the next quarter.

Expected Outcomes

At the quarterly strategic review meetings, the progress of each of the delegated institutional strategic objectives will be measured versus plan through the use of KPIs and milestones. Shortfalls in performance will be discussed and action plans developed to correct the problems encountered. Problems remaining on the action plans for two or more quarters will indicate a serious threat to meeting the long-term strategic objectives and may require the active intervention of the cognizant manager.

Follow-Up

The manager should designate a person to take notes at each of the meetings. Meeting minutes should be generated, reviewed, and distributed to the meeting participants and the management level above the cognizant manager, shortly after each meeting. The minutes should include an executive summary of the key meeting points, the strategic performance scorecards, the performance review templates, and the updated action plan log.

Best Practices

Excellent meetings do not happen by accident. They require a good deal of preparation, follow up, and the use of appropriate meeting tools. By preparing materials in advance, standardizing the meeting communication tools, and having an efficient and effective agenda, the meetings will be more productive and useful to the participants and the review process less onerous. This is the discipline of good performance management. For those who say there isn't enough time to prepare that thoroughly, there always seems to be enough time to correct miscommunications, omissions, and misunderstandings of the leadership team at a later date. My theory is that for every hour that *you* plan, *your team* saves 3 hours in misdirection, miscommunication, and unproductive tangents.

Since the objective of these meetings is to focus on strategy and longer-term objectives, there are several tested methods to put attendees in the right frame of mind to put aside the day-to-day issues that usually distract strategic thinking. The quarterly meeting is often held at an offsite venue to prevent conducting normal business with staff. The use of phones are prohibited during the meetings (a liberal break period is built into the agenda to follow-up on missed calls and important emails and text messages). A premeeting reading assignment on a special topic is given to stimulate innovation and strategic thinking.

While every attendee recognizes the importance of long-term objectives, there is a natural tendency to delay paying attention to these objectives with the expectation of "catching up" sometime in the future. Institutional leaders take advantage of these meetings to develop a sense of urgency to these long-term objectives by setting interim milestones that must be met and focusing on achieving those milestones at every quarterly meeting.

The meeting agenda focuses mainly on management by exception. During each presentation, those strategic KPIs and IOs that are on or ahead of plan can be quickly noted and the presenting manager congratulated. Those objectives that are behind plan can be discussed in detail and potential solutions recommended. Advantage should be taken at the meeting to solicit advice from the leadership team and all other attendees on what has worked for them in solving similar problems that the presenter is currently facing. This will result in a specific action plan to address the shortfall to be reported on at the next quarterly meeting.

An additional focus of the meeting should be external. In addition to reporting out of organizational performance versus plan, each attendee should communicate any strategic intelligence gathered since the last meeting. This could include international technology trends and breakthroughs; changes in leadership and/or strategy in key client organizations; and general market disruptions that affect the organization's strategy. The leadership team should communicate the minutes of the meeting to their direct reports along with specific actions they committed to.

Meeting Tools

To normalize the communication process and help organize the essential information to be presented, there are several meeting tools that are helpful. These include performance scorecards, performance review templates, and the action log.

Strategic Performance Scorecards

Strategic performance scorecards should be created for each management level of the organization. The scorecards contain the specific KPIs and metrics that are aligned with the strategic thrusts. An abridged example is shown below illustrating the level of focus necessary to communicate the strategy.

Strategic Objective	KPI	Metric	Target	Comment
Customer focus	Customer satisfaction	Average score of client satisfaction surveys	>8/10	Output from account management process
Technology leadership	Number of publications	Total number of publications per year	One publication per researcher	Publications in peer-reviewed journals
Building centers of excellence	Growth in research staff	Number of net new hires	25	Departments to reach critical mass
Commercialization of technology	Growth in IP	Number of invention disclosures	20	Approved by the IP committee
Culture of achievement	Staff satisfaction	Average score of staff satisfaction surveys	>8/10	Performance and development scores

Performance Review Templates

Performance Review Templates are an expanded version of the performance score-cards used to present progress being made on the institutional performance objectives developed for each manager at the beginning of the year. The content of each of the templates is taken directly from the completed and approved performance plans. The template lists the strategic and operational institutional performance objectives followed by the scope of the activity, deliverables, milestones to be achieved, progress to date, and comments.

Each manager can fill in the template directly from his IOs, agreed upon at the beginning of the year, and use it to report out on progress at both the quarterly strategic performance management meetings (strategic objectives) and the monthly operational performance management meetings (operational objectives). An abridged example is shown below.

Action Log

During the course of the meeting, several recommendations will be made to each presenter to help bring delinquent items back on track. Keep in mind that the purpose of these meetings and recommendations is to help each leader meet their IOs, and the cognizant manager should facilitate the discussion in a helpful manner.

At the conclusion of the discussion, certain actions will be agreed upon to try. These actions should be documented in the action log. At the next meeting, part of the leader's presentation should be to report on the action taken from the log and whether it helped to advance the progress on the performance objective. The action log commits the leader

Strategic Thrust	Objective (from performance plan)	Scope	Deliverable	Milestones	Progress	Comments
Client focus	Key account plan developed and implemented	Monthly client visits Client engagement meeting	Key account plan Monthly client reports Client engagement report	January Monthly December		
Technology leadership	Publications Strategic alliances	Peer-reviewed publications New alliance	120 Memorandum of understanding	Quarterly submissions Monthly progress report		
Building centers of excellence	Growth in staff	Net new hires	25	Monthly progress report		
Commercialization	Growth in IP	Number of invention disclosures	20			
Culture of achievement	Staff satisfaction	Staff improvement initiatives	Annual increase in score	Monthly progress report		

publicly to initiate agreed upon suggestions to improve on the performance objective. An action log template is shown below.

Date	Action	Lead	Status	Comment

Meeting Minutes

The final tool in the performance management tool kit is the meeting minutes. As stated earlier, the meeting minutes document important news shared (both internal and external), key items discussed, decisions made, acknowledgement of accomplishments, and identification of problem areas and actions to be taken by the next quarterly meeting.

These meeting minutes help share information, decisions made, and actions going forward to those who couldn't make the meeting as well as important stakeholders and managers above and below the performing management unit. Think of all the times that information has been miscommunicated or just plain forgotten from meetings you have attended. The time taken to document the results of the strategic review meeting will make your job easier not harder.

Operational Performance Management

Operational performance management focuses on meeting current yearly commitments to all stakeholders, including internal customers, clients (paying customers), upper management, and staff. As such, operational objectives and KPIs are embedded in operational plans, key account plans, program and project plans, staff development plans, and support division plans. Meeting these operational objectives is both important and immediate and is a major component of performance planning, appraisal, and development process. Therefore, the operational performance management process involves regularly scheduled monthly meetings at each management level focused on early detection and communication of potential problems that might impact meeting the annual operational objectives.

Operational Review Meetings

Rationale

The primary delegated authority for operational performance lies with the senior leadership team. While the leadership team can delegate this authority, each leader is still responsible for the operational performance of the organization and its components.

To ensure the operational performance of his direct reports, each manager should meet formally with each of his direct reports monthly to review performance versus plan. In addition to these formal meetings, informal meetings should be occurring frequently in the course of day-to-day business operations or as a result of specific operational issues.

Schedule

The cognizant manager should conduct operational performance review meetings monthly with his senior leadership team. The meetings should be regularly scheduled to allow sufficient time for the generation and distribution in advance of the operational performance scorecards (see later). The leadership team should meet with their respective management teams and consolidate their teams' input before the meeting. This process should take place throughout the organization.

Attendees

At the operational review meetings, in addition to the cognizant manager and his leadership team, key technical and administrative staff should be invited. Since these meetings take place monthly, this doesn't represent a significant number of overall man hours out of their schedule and can reap significant benefits. These include improved communication and transparency, increased ownership in the plan by key staff, and valuable feedback and problem solving of the plans' shortfalls.

Purpose

The primary purpose of the operational review meeting is to review the progress toward meeting the *short-term* IOs identified in the strategic plan. The focus is entirely on the operational goals. These monthly meetings should be used to identify shortfalls that potentially affect the organizations' strategy and brainstorming to support the affected manager. They should also review progress on the development goals of key staff.

Preparation and Inputs

A week prior to each of the performance review meetings, the operational performance management scorecards are distributed (see meeting tools later). After reviewing their scorecards and meeting with the leaders in their organizational units, each member of the leadership team should prepare a consolidated presentation on progress versus plan on their operational goals. The focus should primarily be on the KPI shortfalls, reasons proposed for the shortfall, and specific actions to be taken in the following month to improve the KPI. Also, each leader should submit in writing to the cognizant manager any topic of concern that should be discussed at the upcoming meeting.

Agenda

The operational review meeting should last no more than an hour and preferably less, depending on the size of the leadership team. This is best achieved by proper preparation and an agenda focused on the KPIs using the traffic-light approach.

Management introduction and overview—5 minutes—The cognizant manager should update his leadership team by providing an overview of his units' operational scorecard, focusing on recognizing performance achievements (green KPIs). Using management by exception, he will be expecting each of his direct reports to be presenting only those areas that are behind plan (yellow and red KPIs).

Leadership team members with operational objectives—10 minutes each—Each member of the leadership team will report on the progress versus plan of their operational objectives using the performance review template (see later). The focus will be on identifying the red flags, causes for the lack of progress, actions taken in past months (see action log later). After each presentation, 5–10 minutes should be allocated for discussion, to include questions and suggestions. This is an opportunity for each leader to obtain the collective insights of the team in problem solving any barriers to meeting objectives.

Open session—10 minutes should be allocated to one or two discussion topics that have been raised in advance. The cognizant manager should choose from the topics submitted by his leadership team before the meeting.

In the beginning, you will find it difficult to accomplish the review in an hour and that's OK. You will find with time and the use of the meeting tools, the meetings will get shorter and the results more meaningful.

Expected Outcomes

These meetings represent an opportunity for the cognizant manager and each of the managers in the organization to regularly review their respective operational performances. Shortfalls in the operational goals should be identified and action plans developed to mitigate the shortfalls. The performance status should be communicated up and down the organization.

Follow-Up

The manager should designate a person to take notes at each of the meetings. Meeting minutes should be generated, reviewed, and distributed to the meeting participants and the management level above the cognizant manager, shortly after each meeting. The minutes should include an executive summary of the key meeting points, the operating performance scorecards, the performance review templates, and the updated action plan log.

Best Practices

As discussed in the section on strategic review meetings, prepare materials in advance, standardize the meeting communication tools, and have an efficient and effective agenda; the meetings will be more productive and useful to the participants and the review process less onerous. It takes some practice (as with everything) to get proficient at conducting review meetings. This is the discipline of good performance management.

Always start and end the meeting on time. This will show respect for the time constraints of your team and allow you to lead by example.

The meeting agenda should focus mainly on management by exception. Those objectives that are on or ahead of plan can be quickly reviewed and the cognizant manager can be congratulated. Those objectives that are behind plan can be discussed in detail and potential solutions recommended. Advantage should be taken to solicit advice from all members of the leadership team on what has worked for them in the subject area and a list of actions to be tried in the upcoming quarter.

Specific actions for each member of the leadership team should be developed to address any KPI shortfalls listed on the action log. Progress on the action plan should be reported on at the next monthly meeting.

Discussion on contentious topics should be time limited and delegated to the action log for further discussion so as not to sabotage the meeting schedule.

Meeting Tools

As discussed in the section on strategic review meetings, to normalize the communication process and help organize the essential information to be presented, there are several meeting tools that are helpful. These include performance scorecards, performance review templates, and the action log.

Operational Performance Scorecards

Operational performance scorecards should be created for each management level of the organization. The scorecards contain the specific KPIs and metrics that are aligned with the operational objectives. An abridged example is shown later illustrating the level of focus necessary to communicate the strategy. You will need to develop specific performance scorecards based on your strategic and operational thrusts.

Operational Objective	KPI	Metric	Target	Comment
Customer focus	Client revenue On-time project delivery	Revenue versus plan Average number of days late	$$/month 0	As per operating plan Level of outliers
Technology leadership	Number of international presentations by research staff	Total number of presentations per year	One presentation per researcher	Targeted conferences
Building centers of excellence	Quality of deliverables	Client satisfaction surveys	>8/10	Month over month improvement
Commercialization of technology	Number of ideas with commercial potential	Five per research unit	25	Approved by the IP committee
Culture of achievement	Staff development	Number of training courses conducted	15	Staff reviews

Performance Review Templates

Performance review templates are an expanded version of the performance scorecards used to present progress being made on the institutional performance objectives developed for each manager at the beginning of the year. The content of each of the templates are taken directly from the completed and approved performance plans. The template lists the strategic and operational institutional performance objectives followed by the scope of the activity, deliverables, milestones to be achieved, progress to date, and comments.

Each manager can fill in the template directly from his IOs, agreed upon at the beginning of the year, and use it to report out on progress at both the quarterly strategic performance management meetings (strategic objectives) and the monthly operational performance management meetings (operational objectives). An abridged example is shown later.

Operational Thrust	Objective (from performance plan)	Scope	Deliverable	Milestones	Progress	Comments
Client focus						
Technology leadership						
Building centers of excellence						
Commercialization						
Culture of achievement						

Action Log

During the course of the meeting, several recommendations will be made to each presenter to help bring delinquent items back on track. Keep in mind that the purpose of these meetings and recommendations are to help each leader meet their IOs, and the cognizant manager should facilitate the discussion in a helpful manner.

At the conclusion of the discussion, certain actions will be agreed upon to try. These actions should be documented in the action log. At the next meeting, part of the leader's presentation should be to report on the action taken from the log and whether it helped advance the progress on the performance objective. The action log commits the leader publicly to initiate agreed upon suggestions to improve on the performance objective. An action log template is shown later.

Date	Action	Lead	Status	Comment

Meeting Minutes

The final tool in the performance management tool kit is the meeting minutes. As stated earlier, the meeting minutes document important news shared (both internal and external), key items discussed, decisions made, acknowledgement of accomplishments, and identification of problem areas and actions to be taken by the next monthly meeting.

These meeting minutes help share information, decisions made, and actions going forward to those who couldn't make the meeting as well as important stakeholders and managers above and below the performing management unit. Think of all the times that information has been miscommunicated or just plain forgotten from meetings you have attended. The time taken to document the results of the strategic review meeting will make your job easier not harder.

☐ Communicating Any Progress Shortfalls and Actions to Be Taken to Correct Them

The fifth and final step in the performance management process is the most critical. If the first four steps are properly conducted, the entire management focus of the organization shifts to celebrating the accomplishment of performance goals and communicating and problem-solving performance shortfalls. Let's review each of the performance management steps again to show the power of this discipline.

In Step 1, the organizational strategy was translated into concrete performance objectives that *must* be accomplished. There must be a high degree of confidence that if the specific performance objectives are met, the strategy will be a success. This translation is easier than it sounds. As mentioned previously, objectives are often chosen for their ability of being met rather than on achieving success of the strategy. More often than not, it's the most difficult objectives that need to be selected. Each IO needs to be challenged to ensure that it is the path to success of the strategy. This process results in a strong alignment of strategy and execution.

In Step 2, metrics were established (KPIs) so that the performance objectives can be measured to ensure that they are being met. Once again, selecting the KPIs is critical. A common mistake some organizations make is to develop and monitor too many indicators. This creates additional, unnecessary work and can distract attention from the key indicators worth monitoring. There should be one or two key indicators that pinpoint the progress of each performance objective. The KPIs need to be SMART. This is an acronym for Specific, Measurable, Attainable, Relevant, and Time bound. In this way, there is no ambiguity in the measurement or its relationship to the performance objective.

In Step 3, the performance objectives are cascaded down and aligned throughout the organization. The best way to accomplish this is to have each manager present his personal performance objectives to his management team. He can then ask each of them to develop their own objectives focused on contributing to meeting his objectives. While this may require several iterations, it is critical to get this step right. Ideally, every manager, and for that matter every individual, should be able to unambiguously state how their performance directly contributes to the IOs of the organization.

10.4

Every manager, and for that matter every individual, should be able to unambiguously state how their performance directly contributes to the IOs of the organization.

In Step 4, quarterly strategic and monthly operational performance management meetings are conducted to monitor whether or not the IOs are being met. The focus of the meetings is to identify and document performance shortfalls and provide support in finding solutions to performance problems. While there is a myriad of discussion topics that could be raised, they should be dealt with at other meetings. These performance management meetings are focused solely on helping managers meet their performance objectives. If the meetings are well planned, conducted in a supportive and open environment, with appropriate follow-up, managers will come to understand how productive they can be.

In the final Step 5, all of the performance shortfalls, both strategic and operational, have been identified, documented in the meeting minutes, and potential solutions developed in the action plan. This information is then communicated both up and down one level in the organization. This process will rapidly develop an open, collaborative culture where meeting performance objectives is paramount. Instead of facing daunting performance challenges on their own, managers can solicit help and advice from other managers to whom the minutes and action plans have been shared.

A couple of words of warning here. It is important to publicly and privately congratulate managers on those performance objectives that are being met. Focusing only on performance shortfalls distorts the overall performance picture and is highly demotivating. If you suspect that a manager is not up to the task at hand despite repeated support from you and fellow managers, it is important that he be spoken to privately and reassigned or replaced. Remember that failing to meet performance objectives is tantamount to not succeeding with your strategy.

It is important to build a culture of trust. Make it clear that if a manager labels a performance objective as green, you expect that performance objective to be met. If a performance objective is labeled yellow, you feel confident that the performance objective will be back on track by the next meeting. And the manager will lose your trust immediately if any of his performance objectives are suddenly labeled red after repeated months of green.

Developing a management team that adheres to this level of openness and trust will require time and your ability to lead by example. Share your performance scorecard at every meeting. Encourage managers to problem solve not only their performance shortfalls but also to offer suggestions on how to help their fellow managers. Weed out those managers who are not team players and are focused only on their personal agenda.

Be sure to always ask whether your team's development objectives are being met and make sure that they are asking about their team as well. Coaching for development is the key to increased performance and commitment from your team. You will see in the next chapter that coaching is the missing link in ensuring high performance.

When the entire organization is acutely aware of the strategic thrusts of the organization; the IOs necessary to successfully execute the strategy; how their personal performance objectives support the IOs as well as their own personal development; and how collectively they can meet their performance objectives; then, the organization is destined to succeed.

☐ Chapter Summary

The execution process involves translating your strategic agenda into actionable objectives and managing to achieve those objectives.

Pitfalls

First, translating strategic thrusts into actionable objectives requires that you pinpoint the critical success factors that ensure the desired outcomes

Second, individual team members have personal aspirations and agendas that do not align 100% with the proposed organizational strategy and tend to go off on tangents

Third, senior leaders incorrectly assume that the translation of strategy and development of a performance management process is not their responsibility because it deals with operations

The Performance Management Process

1. Developing annual IOs that ensure meeting your strategic thrusts

2. Establishing KPIs and milestones to measure whether the IOs are being met

3. Cascading and aligning the IOs to every level of the organization

4. Conducting monthly and quarterly review meetings at each management level to report on progress versus plan

5. Communicating any progress shortfalls and actions to be taken to correct them

☐ References

1. Kotter, John, *Leading Change*, Harvard Business School Press, Boston, MA, 1996.
2. Bossidy, Larry, Charan, Ram, *Execution: The Discipline of Getting Things Done*, Crown Business, New York, 2002.
3. Drucker, Peter, *Managing for Results*, Harper and Row, New York, 1964.
4. Geneen, Harold with Alvin Moscow, *Managing*, Doubleday & Company, New York, 1984.
5. Labovitz, George, Rosansky, Victor, *The Power of Alignment: How Great Companies Stay Centered and Accomplish Extraordinary Things*, John Wiley, New York, 1997.

11

Coaching the Development
Coaching Is the Missing Ingredient in High Performance

In Chapter 3, we listed the universal attributes of coaching in building the trust of team members. Building trust requires *integrity*, behavior that is honest and consistent in words and actions; *empathy*, a genuine concern for the feelings and aspirations of others; and *teaching skills*, the desire to learn the talents and motivations of others and stimulate their self-learning (Figure 11.1).

Despite elegant strategies and formal management processes based on the principles of Chapters 9 and 10, the performance of many organizations fails to meet expectations. Why is this so? From my own experience, I found that the performance of the organizations that I managed dramatically improved when I stopped trying to "manage" direct reports and spent my time coaching them to improve their performance and manage their professional development. I have also seen in my consulting practice a direct relationship between effective coaching and organizational performance by supervisors at all levels. I have come to the conclusion that coaching is the missing ingredient in achieving high performance. If this is true, why isn't it more widely practiced?

11.1

Coaching is the missing ingredient in achieving high performance.

Most organizations recognize the importance of supervising staff and have sophisticated annual performance planning and development programs in place. "Management by objectives (MBO)" has been proven to be one of the more powerful tools in management theory over the past 30–40 years when successfully practiced. Unfortunately, poor execution of the MBO process has many questioning its effectiveness. Most managers and staff oftentimes dread the annual performance review. However, when properly executed, the performance planning and review process is a highly effective way to balance the need for performance (what can you do for the organization?) with the need for personal accomplishment and growth (what can the organization do for you?).

Selecting a compelling destination and a realistic path to getting there

Translating strategy into specific objectives cascaded throughout the organization with confidence

Developing a talented and motivated leadership team capable of executing the objectives

FIGURE 11.1 The Performance Trilogy®.

The following list represents some of the more important reasons why the MBO process is not as effective as it could be. How much does this sound like your organization?

- The majority of the planning effort is spent on filling out forms rather than having meaningful dialog.

- The main focus is tops down on organizational objectives with little input from staff

- The selected performance goals are rarely SMART (specific, measurable, achievable, realistic, and timely)

- Staff aspirations and development goals are given lip service

- Performance goals are rarely reviewed more than once or twice a year

- Development goals are superficial, oftentimes involving taking a training course

- Staff receive little support from management during the year in removing barriers to achieving their goals

In Mihaly Csikszentmihalyi's excellent book *Good Business—Leadership, Flow and the Making of Meaning* [1], he lists the most important factors in determining high performance and staff satisfaction. These are clear goals, instant feedback, a match between difficulty and ability, and a sense of control. To meet these factors, good coaches do their homework. Matching the difficulty of assignments with the ability of staff requires you to spend time getting to know the strengths, weakness, aspirations, and motivations of staff. The best coaches understand that an important part of their leadership responsibilities is placing their staff in positions to succeed, challenging them with meaningful assignments and supporting their development and growth. In a new relationship, the better one understands the aspirations of their staff, the more successful they will be in guiding staff assignments and development.

Staff Development and Renewal

In Chapter 8, we described the seven key processes in leading science-based organizations. The seventh overarching process was staff development and renewal. The three key elements of staff renewal were succession planning, hiring and promoting, and coaching for development and performance. Succession planning is the foundation of organizational development that underpins the entire process [2]. By succession planning, I mean the systematic process of selecting the right staff for each job function and having replacements ready if and when the selected staff are promoted or leave the organization. This is particularly important for leadership positions within the organization, be they technical or managerial.

It is important that you own this process as it will determine who you coach and how you delegate assignments to promote your subordinate's development. In his excellent book, *What to Ask the Person in the Mirror,* by Rob Kaplan [3], he warns that if you haven't identified potential successors for key jobs including your job, it is likely that you are also not delegating sufficiently and are probably a key bottleneck for key decisions. Outstanding people tend to abandon a work environment in which they believe they are not being groomed for greater responsibility through a well-planned series of key job assignments and effective coaching.

At first blush, many managers are afraid that if they groom their replacement, they are jeopardizing their job. Although counterintuitive, it is important to remember that the managers who are most valued in any organization are the ones who develop people. Great companies reward talent developers. Your chances of promotion are magnified if you succeed in your succession plan. I remember telling my management group that my personal goal was to make sure that they all had my job in 3 years. To allay their fears about what happens to me at that point, I pointed out that if each of them were generating the revenues and managing the number of staff I was now, then in 3 years I would be managing five times the current business. My prediction came through.

The purpose of a succession plan is to identify the specific talents and potential of your high performers and matching them to the needs of the business. Do you have candidates that you think can take your place? If not, then this must be a key element of your hiring plan. If the answer is yes, are you spending enough time with them to get to know them better? Are you delegating assignments to them that can challenge their capabilities and coaching them develop their skills?

As a leader and manager, you have the responsibility to create the future for your organization by developing a winning strategy and executing performance goals for you and your staff. At the same time, as a coach, you have the responsibility to design and implement development plans for you and your staff. I am proposing that unlike current performance planning and review processes, you give equal weight to performance and development, as shown in Figures 11.2 and 11.3. I strongly advise that you work with and influence your human resources (HR) department to change the organizational forms to give equal weight to employee development. This will send the right message to employees that they are important, and the organization is serious about helping them grow in their careers. Even if you can't change the forms, make sure that the number of

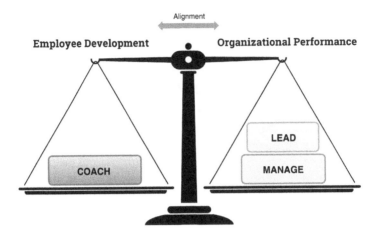

FIGURE 11.2 The performance review process.

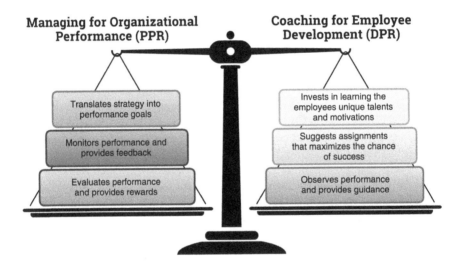

FIGURE 11.3 The performance review process.

development goals roughly equals the performance goals, and as stated previously, start the dialog with development goals first.

☐ Coaching for Development

In coaching for development, the focus is on how the organization can meet the needs and aspirations of the employee. In most performance planning processes, staff development is an afterthought and rarely given much attention. I believe that this is one of the main reasons why organizations fail to reach their goals. In my opinion, the first priority in performance planning should not be determining performance objectives but on employee development. High performance is achieved when staff are self-motivated and give 110% to achieving organizational goals. As stated earlier, staff are motivated by

tailored performance goals that match and challenge their skills, control over the critical success factors, and instant feedback on their
progress. It is nearly impossible to develop tailored
performance goals, unless sufficient time is spent
getting to know the strengths, weakness, aspirations, and motivations of staff. This can only be
accomplished by starting with the needs and aspirations of the staff member [4].

<table>
<tr><td>**11.2**</td></tr>
<tr><td>The first priority in performance planning should not be determin- ing performance objectives, but on employee development.</td></tr>
</table>

Before discussing what an individual's performance goals should be (see coaching
for performance later), begin by discussing staff development goals first. Before you even
look at a form, have several conversations with staff to get to know them personally: their
strengths, weaknesses, likes, dislikes, ambitions, and constraints. Ask their advice on
where they believe they can make the greatest contribution to the organization. Identify
their career goals and realistically evaluate the probability of achieving those goals and
the level of effort and timeframe required. Discuss both on-the-job assignments and
extracurricular activities that would support and advance their career objectives. This
information will be invaluable when trying to align the organizational goals with individual staff expectations.

A big mistake often made in performance plans is to assume that one size fits all.
Many managers justify this position by assuming the only fair process is to treat everyone
equally. This is highly counterproductive. By getting to know staff really well, you can
coach them toward specific assignments that emphasize their strengths and minimize
their weaknesses. Think of the coaching role as similar to a talent agent in the entertain-
ment industry. Great talent agents know their clients' strengths and recommend only
those roles where they know their clients can perform well.

Presented later are a few illustrative examples.

Allow staff to try branching out into new areas with some level of security. As an
example, one of my high-performing senior scientists was interested in a group leader
management position but wasn't sure whether this was a right career path for him. We
both agreed to give it a try and for him to assume the role and responsibility of the group
leader position. This trial assignment was communicated broadly with the understanding that he had his choice to continue in the group leader role or return to the staff
scientist/project manager role. At the end the year, having performed the group leader
role admirably, he decided that he enjoyed technical project work much more and seam-
lessly moved back into a project leader role with no loss of face.

Exploit exceptional strengths for the benefit of both the organization and individual. As another example, one of my senior staff had a brilliant mind and extraordinary
skill in experimental design, scientific interpretation of research, and lucid report writing. However, his organizational and financial management skills prevented him from
advancing in the organization where managing projects was a promotional requirement.
We developed a performance plan for him focused on the role of principal investigator on
multiple organizational projects, a role in which he excelled, and the projects benefited.
By pairing him up with younger project managers, they had the opportunity to learn and
be mentored in sound fundamentals of science. In time, he rose to the highest level on
the technical ladder, research leader.

Have the courage to allow staff to change careers. Sometimes, staff performance
deteriorates because staff feel stuck in a job that does not suit them despite their academic
training and degree. They discover hidden talents after having worked in an organizational setting for a while and want to branch out and change careers. Instead of thinking

that the development cost of a career change is too expensive for your organization, think of it as a good investment in an individual who has discovered not only his or her talent but also the motivation to excel. From personal experience, this can be an excellent return on investment. One example is a financial manager/contracting officer with an MBA who was stalled in his current position. He excelled at negotiations and building client relationships and preferred dealing with clients rather than back office work. I offered him a trial assignment for 1 year in marketing and business development in which he excelled. He went on to have many successful and enjoyable years in that role.

In another example, a very organized scientist with a high level of emotional intelligence, social skills, and maturity applied for a position in HR despite having no background or education in that profession. We took a chance and gave her the assignment based on her talent as opposed to experience with a development goal of acquiring the necessary HR skills through training. She went on to be an outstanding HR professional and eventually became HR manager of the organization.

These examples are meant to illustrate the importance of investing in people; getting to know the talent and motivation of your staff and putting them in assignments where they can succeed.

☐ Coaching for Performance

In performance coaching, the focus is on how the individual staff member can contribute to the goals of the organization. Based on my experience in evaluating the performance planning process in a dozen science and technology organizations in the United States, Latin America, Europe and the Middle East, the process goes something like this. The supervising manager establishes several performance goals for his/her staff based on the goals passed down to him/her in company forms fit for that purpose. These performance plans are sent to the staff member who is asked to review the performance objective, list a few personal development objectives, and then sign up for the year.

On average, the amount of face time that the supervisors spend with each reporting staff member is about 30–60 minutes. In the best case, the staff member agrees with the performance goals. In many cases however, there is some disagreement or ambiguity with the stated goals, and the staff member is reluctantly forced to sign the performance plan anyway. There is little dialog at this time to resolve these disagreements, leading to reluctant compliance on the part of the staff member. Not a situation that results in 110% commitment.

Once the plans are finished, they are often "put on the shelf" and not looked at until the end of the year. In some cases, there is a cursory review midyear. I call this absentee management. When the time comes for the annual end of the year review, the manager has little information on the performance of the staff member versus plan and little concern for the adjustments that needed to be made during the year as a result of changes in priorities. With ambiguous goals and lack of feedback, it's no wonder that performance reviews are dreaded by some staff. One of my consulting colleagues and an expert in HR jokingly remarked once that, with absentee managers, staff would have a hard time picking them out of a police lineup!

In the other extreme, micromanagers not only tell staff what to do, but also how to do it (usually their way). They are constantly and frequently checking up on progress and are quick with their criticisms. This constant negative feedback and loss of control make for a highly negative work environment and low retention rates. Most talented staff don't leave organizations, but they leave their immediate supervisors.

> **11.3**
>
> Most talented staff don't leave organizations, but they leave their immediate supervisors.

There is a better way!

Imagine entering a performance-planning meeting with staff having already spent considerable time on their development plan. You are armed with information on particular strengths of the staff member and what truly motivates them in their job. You both have a clear understanding of the staff members' career aspirations and the steps necessary to achieve them. You also know the staff members' weaknesses and job functions they dislike that may impede their performance. If you have invested the time to get to really know the staff member and have already agreed on a development plan, you enter the performance planning process with an increased level of trust, so essential for a good coach [5,6].

Before meeting with a staff member during the performance planning process, provide as much information as possible ahead of time on the performance expectations of the organization for the upcoming year, your performance goals, and what you expect to delegate to the staff member. Along with the materials, ask the staff member to recommend areas where he/she can make a contribution to meeting organizational goals above and beyond the obvious in his/her job description. Suggest areas yourself based on your knowledge of the staff's strengths and unique talents.

With this level of preparation, the performance-planning meeting can be highly productive. You both enter the meeting with all the information needed to balance the needs of the organization with the aspirations of the staff member. You can then focus on a meaningful dialog by asking the following questions of the staff member. Where do you think you can make the greatest contribution to the organization? What assignments really excite you? What activities do you consider a chore? What are the barriers you foresee during execution? What support would you like to see from me? By discussing performance in the context of the staff members' personal preferences, you are now coaching instead of managing. This process will lead to an optimum performance plan that meets both the needs of the organization and the staff member.

Once the plan is generated, this becomes the road map for the year for both of you. As discussed earlier, staff satisfaction comes from clear goals, instant feedback, challenging but realistic assignments matched to the talent level, and personal control over work assignments. Instead of waiting until the middle or end of the year, plan regular meetings once a month to review progress on the plan. During these meetings, congratulate the staff member on those assignments that are going well; for those assignments that are not going well, ask for reasons why and what he/she plans on doing about it; and most important, always ask if there is anything you can do to help. By meeting regularly and asking such questions, instead of being resented as an absentee manager or a micromanager, you will be appreciated and trusted as a performance coach. At year-end, instead of dreading the performance evaluation, you can write up a thorough report with no surprises. Better still, the staff member can easily draft a performance report for your signature.

> **11.4**
>
> Staff satisfaction comes from clear goals; instant feedback; challenging but realistic assignments matched to the talent level; and personal control over work assignments.

If what I am proposing is the secret to high-performing staff, why isn't it more widely practiced? First and foremost, it boils down to one's limited awareness of sustainable leadership and what it takes to succeed. Some managers, although they wouldn't admit it publicly, consider staff as just paid resources to be exploited for their own benefit. Indeed, in the short run, one can be successful by managing "upward" with an excellent strategy and well-managed execution with little regard to staff. Success in the long run however is sustained by coaching "downward" with an increased awareness that success is achieved through talented staff that are motivated and aligned with organizational goals.

Even with this increased awareness, the biggest objection I hear to implementing the earlier approach is that there just isn't enough time in the day to spend on coaching staff. Nothing could be further from the truth. Let's take an example. You manage a technical group or senior leadership team with ten direct reports. If you committed to spending 2 hours/month of your valuable time focused on performance and development coaching for each staff member (1 hour/month of quality face time and 1 hour/month for preparation and follow-up), this would represent 20 hours/month or a little over 12% of your time. I suspect that you are spending more than that fighting fires each month, many of which could have been prevented with the earlier approach. Great coaches understand the importance of "working for" their staff to help them achieve their individual goals and, as a result, organizational performance.

11.5

Great coaches understand the importance of "working for" their staff to help them achieve their individual goals and, as a result, organizational performance.

☐ Balancing Organizational and Staff Needs— The True Meaning of Alignment

Unfortunately, even if you begin to practice development and performance coaching described earlier, there will always be difficult staff to deal with. One of the biggest mistakes that managers can make is not dealing with problem staff. One way to frame the discussion and help make decisions is to assume that most problems result from a misalignment between the aspirations of the staff member and the needs of the organization. Everyone has the talent and motivation to perform in areas of personal and professional interest. The problem lies in whether there is sufficient alignment of their interests with organizational goals. During the performance planning process, it is rare that personal interests align 100% with organizational goals, and I believe that getting to at least 70% is essential to ensure a high level of performance. One of your challenges is to get to 70% alignment during the planning process.

A staff member's inability to perform can be due to either a lack of talent or motivation. I can think of three good examples I've experienced. In the first case, an individual you know to be talented underperforms due to lack of self-confidence and fear of failure. This is a good example of where good coaching can dramatically improve performance. Try giving encouragement along with frequent positive feedback, and increasingly more challenging assignments with a safety net to avoid fear of failure. This will build self-confidence and dramatically improve both performance and staff satisfaction. If you are willing to invest the time, this is a highly rewarding experience.

A second, more problematic example is an individual who has an inflated view of his/her talent and unrealistic expectations for career advancement. The challenge here is to spend the time during the development plan to point out the specific skills that would need to be developed to achieve his/her personal ambition. It is also important to temper the staff member's desire to take on assignments that you know they are not ready for. Once again, a good coach is like a talent agent who chooses roles that are right for the person. I made the mistake once of telling a staff member that I didn't think he or she had the talent to achieve his or her career goal and lost that person's trust right from the start.

There will sometimes be the need, whether due to lack of talent and/or motivation, to recommend that an individual would be happier in a less challenging role or different organization. Putting off such a difficult conversation is bad for both the individual and the organization.

The greatest return on investment of your time is getting to know your staff; what they are good at; and what motivates them. This investment will pay immediate dividends and take less and less time each year as you get to know your staff better. There is no greater path to high performance than coaching a talented and motivated staff.

☐ Chapter Summary

"MBO" has been proven to be one of the more powerful tools in management theory over the past 30–40 years when successfully practiced.

The three key elements of staff renewal are succession planning, hiring and promoting, and coaching for development and performance.

Succession planning is the systematic process of selecting the right staff for each job function and having replacements ready if and when the selected staff are promoted or leave the organization.

Staff are motivated by tailored performance goals that match and challenge their skills, control over the critical success factors, and instant feedback on their progress. In coaching for development, the focus is on how the organization can meet the needs and aspirations of the employee.

In performance coaching, the focus is on how the individual staff member can contribute to the goals of the organization. If you have invested the time to get to really know the staff member and have already agreed on a development plan, you enter the performance planning process with an increased level of trust, so essential for a good coach.

Work with and influence your HR department to change the organizational MBO forms to give equal weight to employee development. This will send the right message to employees that they are important, and the organization is serious about helping them grow in their careers.

☐ References

1. Csikszentmihalyi, Mihaly, *Good Business, Leadership, Flow and the Making of Meaning*, Penguin Putnam, New York, 2003.
2. Charan, Ram, *Leaders at All Levels*, John Wiley, New York, 2008.

3. Kaplan, Ron, *What to Ask the Person in the Mirror*, Harvard Business Review Press, Boston, MA, 2011.
4. Wall, Bob, *Coaching for Emotional Intelligence*, Amacom, New York, 2006.
5. Fournies, Ferdinand F., *Coaching for Improved Work Performance*, McGrawHill, New York, 2000.
6. Clutterbuck, David, *Coaching the Team at Work*, Nicholas Brealey International, New York, 2007.

Putting It All Together
Leading Is a Team Sport

In this chapter, I will summarize the essence of the Performance Trilogy® leadership model and reemphasize the key aspects of putting the model to work for you, whether it be at the personal, managerial, or executive level. Table 12.1 is an overview of the key principles discussed throughout the book. As I stated in Chapter 1, there are many more subprocesses, objectives, and leadership attributes, that have to be managed. I am confident however that the three processes represented by the Performance Trilogy form the fundamentals from which one can ensure any successful leadership initiative.

☐ Executing the Performance Trilogy®

I do want to add a little more complexity to the model in two ways. First, it is highly unusual to find leaders that possess a high degree of competence in all three roles: leader, manager, and coach. Usually, the dominant attributes that you possess will tend toward competence in one or two of the roles. When first starting out in one's career, the most common role and responsibility and the dominant skill set is as a manager, executing the vision and strategy of those higher in the organization. It is not unusual however that some professionals possess the qualities of a leader and lack management skills. This was certainly my case. I have found that the least likely skill set is that of the coach. As mentioned previously, it is counterintuitive for highly ambitious people to view developing others as the path to success. Their mindset of their staff is to utilize them as resources to get what they need. The wisdom of coaching usually comes later in one's career.

The good news is that you don't have to have all the attributes to lead, manage, and coach. The important thing is to know what your strengths are (remember the discussion on self-awareness) and surround yourself with a team that possesses complimentary skills and attributes. Play to your strengths, be aware of your weaknesses and shore them up with your team. Subjugate your ego and recognize what each of your team members

TABLE 12.1 Summary of the Performance Trilogy

Performance Trilogy	Strategy	Execution	Leadership
Objectives	Develop a winning strategy and getting buy-in from participants	Translating strategy into performance objectives that are actively managed	Selecting and developing the leaders to execute the strategy
Dominant role	Leader	Manager	Coach
Leadership perspective	Lead from the front	Lead from the middle	Lead from the rear
Key leadership attributes	Imagination	Conscientiousness	Empathy
	Courage	Productivity	Integrity
	Persuasiveness	Discernment	Teaching skills
Desired emotions from participants	Inspire faith	Build confidence	Engender trust
Personal leadership	Who am I and where am I going?	How am I going to get there?	Who do I want to become?
Managerial leadership	Team vision and strategy	Action plan	Leadership development
Executive leadership	Corporate strategic planning	Performance management	Talent management

brings to the leadership table; pay attention to them; and allow them to play to their strengths. In Figure 12.1, the key roles and responsibilities of the leader, manager, and coach are presented as a universal leadership model.

I know from personal experience how valuable it is having a team that is aware of each other's strengths, honors them, and allows them to take the lead when needed. It took me way too long to learn this lesson and I hope that my lessons learned can help you get there faster than I did. I came to realize that, earlier in my career, my major strength was in strategy. I was always focused on the big picture and quickly became impatient with the day-to-day activities needed to execute. As a result, I gravitated toward taking over and leading troubled organizations in need of transformational change. I was successful with turning around those organizations but when they stabilized, it was time for me to move on and let a good manager run those businesses.

About midcareer, I found myself needing to manage a much larger organization and needed to change my mindset to that of a manager as well as a leader. I worked hard at improving my skills at conscientiousness, productivity, and discernment and also surrounded myself with highly competent managers from whom I learned a lot. It was also at this time that I found my decision process changed. I began to realize that my

FIGURE 12.1 The universal leadership model.

decisions improved markedly when I factored in the needs of the staff as well as my desired outcome.

While it was a gradual process, I developed a new mindset about the definition of success. I wanted to build and grow a successful organization that people really wanted to work at. It had to first of all be successful, which would require hard work, but more importantly, it had to be a place where people had fun and knew that they would be recognized for their accomplishments. As a result, my focus turned toward becoming a good coach for my high performers. I was surprised at how easy the transition was. Seventy percent of the change came from a change in my mindset. Once my staff saw that my intent was to make them successful, their trust level increased dramatically. The 30% of mechanics I had to learn over time.

The lesson learned for me is to play to your strengths but always be willing to grow based on new situations. As a lifelong learner, my motto is "better than yesterday, not as good as tomorrow."

The second point that I would like to emphasize is that while the book presents the Performance Trilogy in sequence so as to understand the principles of each phase, it actually is a iterative process. When developing a winning strategy, it is important to factor in all of the issues that might arise in executing that strategy. There may be a need to coach some of your leadership team at the strategy phase as well. During the execution phase, you often will have to circle back to your strategy to make sure that it is still winnable based on new data obtained. Coaching is constantly taking place during the execution phase as well. As mentioned in Chapter 7, the best form of training is on-the-job training, and it is highly cost effective and productive to provide for staff training and coaching on every project. And finally, good coaching takes place throughout the Performance Trilogy, making sure that key staff are properly aligned, motivated, trained, and rewarded.

☐ Balancing Performance and Development

One of the most important takeaways from this book is that there is no magic bullet to leadership. Too many books and articles promote such simple solutions. "Take care of your employees and they will take care of your business." "The secret to success is to super-please your customers." "Be an authentic leader." The secret to any successful enterprise is to balance all of these needs and manage the conflicts that inevitably occur. I have mentioned in previous chapters the importance of gaining trust as a critical factor in any leader's success. The best illustration I have found on the foundation of trust that matches the universal model of leadership presented in the Performance Trilogy is shown in Figure 12.2 and is the title of Shaw's book, *Trust in the Balance* [1].

As a leader, the first imperative is to achieve results. In a business environment, this requires the strategy of a leader and the execution of a manager. Trust and credibility comes first and foremost from following through personal and business commitments. No amount of time and attention paid on "people skills" will overcome the lack of success in achieving results. A very close colleague of mine amusingly described the difference between personal trust and business trust this way. "I very much like my future son-in-law. I trust him well enough to marry my daughter, but I would never let him run my business!"

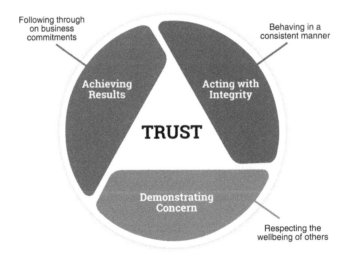

FIGURE 12.2 Building a culture of trust.

The second imperative is to act with integrity. Any form of disingenuous behavior will always diminish the trust level of the people you work with and deal with. So, from this perspective, as you know, I am a big fan of Bill George's philosophy espoused in his book, *Authentic Leadership* [2]. Living your values, behaving in a consistent manner, and balancing the need of multiple stakeholders, even if it means disappointing some, will engender a great deal of trust in you as a leader.

And finally, demonstrating concern for the well-being of others requires the attributes of a good coach. When developing a strategy, sufficient thought needs to be given to what the benefits will be, not only for you, but also for those followers who took considerable risk in supporting you. When executing the strategy, ask what you can do to advance your staff's aspirations and career goals in addition to demanding that they meet the organizations' goals. And finally, invest your time in getting to know your staff well enough to provide advice and counsel when needed.

Achieving this balance is not easy. It takes a great deal of dedication and hard work as well as an unselfish mindset. This is the reason why there is a dearth of great leaders.

☐ It's All about Talent Management

So, within the framework of business success and hard decision making, now we can conclude with the key takeaway from the book. Leading the strategy and managing the execution are your responsibilities, but they are performed by your staff. How you select, train, coach, and reward your staff plays a critical role in whether you are successful as a leader. To summarize the elements presented in previous chapters:

- Select for talent first, motivation second, and experience third

- Set high standards starting with yourself and reward accordingly

- Get to know the unique talents of your staff and put them in roles where their talents can support their success

- Monitor the performance of your staff often in a supportive way

- Coach up your high performers and challenge them to achieve their full potential

- Place your marginal performers in appropriate roles or outplace them

- Spend *equal* time during the performance planning and review process on the needs of the staff member as well as the organization and work toward obtaining 70% alignment between the two.

I would like to end by focusing on the last bullet point as I believe that it represents an original contribution to the management literature. When starting the performance planning and development process for the coming year, start first with the development process. Spend considerable time getting to know the talents, aspirations, and career goals of the staff member. Work with a staff member to come up with development goals unique to that individual and supportive of your organizational goals. These could include a development assignment on a task force, mentoring, on-the-job training, or a good book recommendation. Commit to reviewing these development goals like any good coach would. Provide encouragement when needed.

Once these development planning meeting or meetings have taken place, I am confident that the performance planning meeting where you discuss what the staff member can do for the organization will go much smoother. I have found that the staff member pretty much can come to the meeting with performance goals that you will be happy with requiring little modification. Plan on reviewing both performance goals and development goals monthly at a minimum and give them equal time.

You will be surprised that, by adopting this approach, the amount of time that you spend on the annual performance planning and development process will not increase. Instead of being overwhelmed at the end of the year with performance reviews and dealing with missed assignments and miscommunications, you will find that you are pretty much on top of all of the issues, having spent the time every month coaching your staff. You will also see that the staff satisfaction of your group increases dramatically.

I have given you a lot to think about. Adopting the Performance Trilogy will require a change in mindset, learning new skills, and lots of practice. I am confident that you will find the effort worthwhile.

☐ References

1. Shaw, Robert B., *Trust in the Balance, Building Successful Organizations on Results, Integrity, and Concern*, Jossey-Bass, San Francisco, CA, 1995.
2. George, Bill, *Authentic Leadership*, Jossey-Bass, New York, 2003.

INDEX

A

Academic literature, 24
Authentic leadership, 191

B

Beta test project, 81
Building confidence, in execution, 14–16
Business
    development, 106–107
        in technology-based business, 82
    literature, 1
    management in technology-based business, 84

C

Client feedback, 121–122
Client-relationship management (CRM), 88, 89
Client strategy, 91, 146–147
Coaching attributes, 44–48
Coaching development, 177–185
Coaching—gaining trust through development, 16–18
Coleman, Daniel, 31
Communications strategy, 145–146
Community strategy, 147–148
Compelling strategy, 151
Competitor analysis, 122–124
Conscientiousness, 40–41
CRM. *See* Client-relationship management (CRM)
Csikszentmihalyi, Mihaly, 178
Customer service model, 84–85

D

Destination
    achievable, 12–13
    beneficial, 13–14
    desirable, 12
Development planning and review process (DPR), 5
Difficulty of assignment, 25–26
Discernment, 43–44
DPR. *See* Development planning and review process (DPR)

E

Emotional intelligence, 31
Empathy, 46–47
Estimating magnitude of challenge, 25–27
Evaluating performance, 72–73
Execution
    building confidence in, 14–16
    managing, 151–175
    Performance Trilogy®, 187–190
    project leader's role in, 92–93
    turning intentions into actions, 4–5

F

Financial management, 105–106
Financial strategy, 141–142
Foster self-learning, 47–48

H

Huxley, Thomas, 43

I

Important activities
    and nonurgent activities, 65–66
    and urgent activities, 65
Integrity, 45–46

K

Knowledge Management System, 117

L

Leadership
    attributes, 35–48
    challenge, 27–33
    framework, 9–18
    personal, 23–33
    project, 79–97
    quarterback of performance, 5–6
Leading—inspiring faith in strategy, 11

M

Management
    attributes, 40–44
    literature, 10
    strategy, 91
Management by objectives (MBO) process, 177
Managerial identity/style, 57–60
Managing—building confidence in execution, 14–16
Market analysis, 126–129
Market-driven company, in technology-based
        business, 86
Market strategy, 134–136
Market survey project, 81
MBO process. *See* Management by objectives (MBO)
        process

N

National Institute of Health's (NIH) guide, 61
NIH guide. *See* National Institute of Health's (NIH)
        guide
Nonimportant activities
    and nonurgent activities, 66–68
    and urgent activities, 66

O

Operational performance management, 156, 162, 169
Operational review meetings, 169–174
Operative strategy, 4

P

Paper pusher, 54
Performance management process, 154
Performance planning and review process (PPR), 5
Performance-planning meeting, 183
Performance Trilogy®, 1–6
Personal leadership development, 32
Persuasiveness—ability to influence, 38–39
Pilot-scale project, 81
"Player-coach" role, 24
PPR. *See* Performance planning and review process
        (PPR)
Product development, 108–110
Product strategy, 137–140
Project leadership, 79–97
Project management, 107–108

R

Regulatory strategy, 143–146
Resource strategy, 140–143

S

Science-based business, 143
Self-assessment, 118–121
Self-awareness, importance of, 29–33
Service delivery, 89
Staff development
    and renewal, 110–112, 179–180
    in technology-based business, 83–84
Staff strategy, 91, 140–141
Stakeholder
    requirements, 129–131
    strategy, 146–148
Strategy
    blueprint of performance, 3–4
    leading, 115–148
    performance management, 156, 162
    planning process, 103–104, 115–118
    project leader's role in, 92
    review meetings, 162–169
    synthesis, 131–134
Succession planning, 179
Sun Tzu, 24
Supervision, 63–76
Supplier strategy, 147

T

Technology-based business
    business development in, 82
    business management in, 84
    customer service model, 84–85
    market-driven company, 86
    staff development in, 83–84
    traditional product-driven company, 86
Technology trends, 124–126
Traditional product-driven company,
        in technology-based
        business, 86

W

Webster, 10, 92
Winning strategy, 11, 115–116, 135